What Philosophy Does

What Philosophy Does

Richard Lindley (editor)
Roger Fellows
Graham Macdonald

Open Books
London

First published in 1978 by Open Books Publishing Limited,
11 Goodwin's Court, London WC2N 4LB

Hardback ISBN 0 7291 0137 1

Paperback ISBN 0 7291 0142 8

Text set in 11/12pt Photon Imprint, printed by photolithography, and bound in Great Britain at The Pitman Press, Bath

Contents

For Errollyn, Sue and Cindy

Introduction

Philosophy is widely regarded as a very academic subject, because it can seem very remote from people's ordinary lives. We think this is unfortunate, because philosophy can be of use and interest to a much wider group of people than select members of the academic community. Of course there are many different branches of philosophy, some of which seem more obviously relevant than others. Also, one can study the subject at a number of levels. In this respect philosophy is similar to the natural sciences, which most people study, albeit at an elementary level, at school. Because philosophy is not often studied outside universities, even *introductory* philosophy books tend to be too difficult or too specialised.

Many people who think that they might be interested in philosophy are put off by reading introductions to the subject which are too difficult to understand, and which don't really try to show why the problems discussed by philosophers might be important for ordinary people. In this book we have sought to avoid these pitfalls by using as little philosophical jargon as possible, and attempting to show how philosophy can contribute to solving problems which confront most of us in one way or another. Amongst the topics we discuss here you will find the questions of whether abortion should be available on demand, whether we should believe the claims of Uri Geller that he can bend metal objects by psychic power alone, and whether animals can be said to possess language.

Despite first appearances these questions are all, in their way, philosophical. Indeed we have chosen them, not just because they are topical matters for the general public, but because they are subjects which are of great concern to contemporary philosophers. Our book differs from many introductory philosophy books in that it is in no sense historical. Many introductions to philosophy take the form of a potted history of the subject. Such an approach undoubtedly has its merits, but we think it is more exciting for many people to go straight to issues and arguments which are of current philosophical concern. In any case there is no shortage of good histories of philosophy.

Naturally, the approach adopted here is not without its own dangers. There is much controversy amongst philosophers over what exactly philosophy is, or should be. Because of this, any introduction to the subject which is not simply a record of what philosophers have said is bound to be controversial. We don't, however, try to define exactly what philosophy is or should be; instead, we take a number of topical issues, and show how the different branches of philosophy can be brought in to help solve them.

A second danger arises because in trying to keep our arguments simplified it is not possible to do full justice to the complexities of the issues we discuss. To do this we would have had to write quite a different book – a book which we think would be inappropriate as an introduction to the subject. Nevertheless, we do stick by the views we put forward, even though we have not here defended them against all objections.

The book does become progressively more difficult. This is partly because the subjects discussed towards the end of the book are more abstract and theoretical than those in the earlier sections, and partly because we wish to give the reader scope to use, in dealing with the later sections, skills picked up in the earlier sections. Nevertheless, we would hope that someone who was particularly interested in, say, philosophy of science, could benefit from reading our chapter on this subject without reading the earlier ones.

The authors all teach philosophy in a university department

which combines the study of philosophy with that of sociology, psychology and literature. Apart from teaching within this department we also teach philosophy to other students in the university who are taking degrees in management studies, environmental science, psychology, and peace studies. We are grateful to these students for giving us practice at trying to get philosophy across to non-philosophers.

We have all contributed, at least through discussion, to each of the chapters. We would like to thank all those who have encouraged us in this project, and in particular Henry Hardy for his patient, thorough and constructive criticisms of earlier drafts of this book.

1

Social philosophy

> The conventional view that philosophical discussions are quite remote from having any practical upshot, such as prevention of suffering or loss of life, has very little to be said for it.
>
> (Glover 1977, p. 16)[1]

WHAT IS SOCIAL PHILOSOPHY?

Is abortion right or wrong? Should severely handicapped infants be allowed to die? Ought voluntary euthanasia to be legalised? Should mentally ill people be forced to undergo treatment in mental hospitals? What, if anything, is wrong with private health care for those who can afford it? Should we have comprehensive education for all children? Is it ever right to sacrifice individual liberty for greater equality?

These are important questions which affect us all whether we think about them or not. So it is a matter of concern that there is so much unclear and dogmatic thinking about them. In the last few years philosophers have started to turn their attention to questions such as these. It is not true, as is sometimes suggested, that philosophers are never concerned with topical, down-to-earth problems. Indeed, one of the prime tasks of contemporary social philosophy is to face up to just this kind of issue. Social philosophers try to disentangle the tangles and to produce coherent answers which are also true to what we most value. Social philosophy also deals with more abstract issues such as whether a society is anything more than the individuals who live in it, and whether social sciences such as sociology and economics are really scientific. These issues are also important, but their importance is not so immediately obvious to someone who has not already done some philosophy. Consequently, in this chapter we shall concentrate on how philosophy can help to answer questions such as those in the opening paragraph. Many philosophical problems arise when we reflect on common knowledge and common sense, and it is with common knowledge and common sense that we begin.

CONTROL OVER LIFE AND DEATH

Some of the most important scientific advances in the last hundred years have been in the field of medicine. We are now able to prolong people's lives enormously. We can also kill people painlessly with drugs. We can artificially prevent the conception of embryos, and we can abort unwanted foetuses without a significant risk to the mother's health. These advances have dramatically brought to the fore a number of extremely complex moral problems about our control over life and death. Natural and social scientists can, with the aid of their technical expertise, inform us of the likely consequences of various actions. But they cannot answer for us the basic moral questions which sooner or later inevitably confront both lawmakers and ordinary citizens alike. So we shall start by discussing some of the moral problems arising out of our ability to interfere with natural processes of life and death, looking in particular at the problems of contraception, abortion, infanticide and euthanasia.

CONTRACEPTION

There are many different ways of avoiding the conception of unwanted babies. Some of these methods, such as the use of the 'safe period', coitus interruptus and simple abstinence, do not require the use of artificial devices. Even such a notable opponent of contraception as the Roman Catholic Church maintains that these methods are acceptable. The Catholics contrast these natural methods with artificial methods, which they condemn. These include the sheath or condom, the Dutch Cap and 'the pill'.

Arguments against the use of all artificial contraceptives often rest on the belief that their use is against the will of God. It is argued that since the natural processes of reproduction arise from God's will, if someone interferes with them he is defying the will of God and therefore sinning.

In order to counter arguments which claim that a certain type of action is wrong simply because it defies the will of God you would either have to argue that in fact it is not against His will, or that it is unreasonable to believe that such a God exists, or that if He does exist, we should not pay any heed to His will. Such an argument would take us into the fields of theology and philosophy of religion, which lie beyond the scope of this chapter. Fortunately for our purposes, the God most believers in this country believe in is not a whimsical or unreasonable God. So we can legitimately ask of a practice which God allegedly forbids or condones why He would forbid or condone it. If there is a God in whose image we are created we might assume that the reasons which He has for willing this or that human action ought to be acceptable to us. Consequently we shall concentrate on arguments which don't rely *directly* on a claim that a certain practice is right or wrong just because God says it is.

An argument against artificial contraception which on the face of it is independent of these religious issues is that artificial contraception is wrong because it is unnatural. This sort of argument is used with alarming frequency, especially in the area of sexual ethics. Let us look at it.

The argument that artificial contraception is wrong simply because it is unnatural is based on the view that, in general, whatever is unnatural is wrong. But this would mean that the use of all sorts of things which benefit mankind and don't harm anyone is wrong. Washing machines and washing powders are made of synthetic materials. They are not 'natural'. If it's wrong to use things which aren't found naturally, then it's wrong to use washing machines. It would also be immoral to wear synthetic-fibred clothes. But why should anyone think that wearing nylon shirts is immoral? You might not like the way they look or feel. But that is another matter altogether.

Somebody who upheld the 'unnatural therefore wrong' position on contraception would doubtless say that the comments about washing machines and nylon shirts miss the point. This may be true. But it doesn't invalidate the criticism of the view that what is unnatural is wrong; it simply shows that those who claim that artificial contraception is wrong because it is unnatural are really being motivated by something other than the belief that what is unnatural is wrong.

What usually underlies the claim that artificial contraception is wrong simply because it is unnatural is a belief that sex has a purpose which is thwarted by using contraceptives. People frequently say that our reproductive systems are for reproducing and we should not abuse them. It is true that if everybody practised contraception throughout their years of sexual maturity the human race would die out, and this might be a bad thing. However this new argument is far too extreme. If it is to be effective against the use of artificial contraceptives it would end up ruling out even the so-called 'natural' methods of birth control. It would also rule out as immoral sexual encounters between homosexuals, between infertile people, and all sex for pregnant women. It would exclude any such encounter, irrespective of the pleasures it might bring, and even if it does nobody any harm.

The argument from unnaturalness, which leads to the absurd consequences mentioned above, claims that there is something wrong in interfering with bodily functions, simply because it is unnatural. We must not confuse this dubious claim with the justfiable warning that we should be very careful before we interfere with our bodily systems. Very often people have tried out drugs on people after insufficient research has been done into the side-effects of the drug. The most notorious recent case in Britain is thalidomide. Thalidomide is a drug which was administered to pregnant women as a tranquilliser. Pregnant women often undergo very rapid mood changes. They can feel tense and depressed. Clearly this is unpleasant. The fact that thalidomide could relieve these unpleasant conditions was a good reason for wishing to use it. However, the effects of this 'unnatural' drug were disastrous. It damaged many foetuses, and so produced many severely handicapped children.

But what is wrong with thalidomide is not that it is 'unnatural', but rather that it produces handicapped children. Similarly, it is arguable that 'the pill' can have unpleasant side-effects, even that it can kill the woman who takes it. This provides a reason for thinking twice about using the pill. If the pill does become perfectly safe for women, then this reason will no longer apply, even though the pill will be no less artificial. In fact, as things stand now, the risks to women from medically supervised use of the pill are so slight, and the benefits can be so great, that the argument from the possible dangers to the women who use it is much less powerful than it used to be.

Having looked at a couple of the arguments against contraception, let us now look at its central justification. The main reason for being in favour of contraception is that it benefits people, and apart from the possibility of harmful side-effects, it doesn't harm anyone. A woman who has the benefit of effective contraception is able to make love when she wants to without the fear of pregnancy. On the other hand, the use of contraceptives need not prevent her from conceiving a child later. In this way contraception differs from currently available forms of sterilisation, which are usually irreversible.

As bearing children is a far graver matter than simply fathering them, the development of contraception has provided more immediate benefits for women than for men. But clearly the benefits apply to both sexes, because the conception of unwanted babies usually causes distress, and very often hardship, to both parties, and in an over-populated world, to the general public.

The main argument in defence of contraception is that it gives people the benefit of greater freedom without harming anyone. To argue against this you would have to show either that making contraceptives available to people does not increase their freedom, or that freedom is not always a benefit, or that contraception does harm people.

Most opposition to this argument comes from people who deplore the arrival of the so-called 'permissive society'. One not infrequently hears the following sort of argument: 'Contraception is one of the curses of the twentieth century. Since contraception has been available, the young sleep around with any casual acquaintance. Sex has become cheapened. Married people are unfaithful to each other, and the divorce rate is higher than it's ever been.'

This argument is introduced as an argument against contraception. The implication is that the use of contraceptives is harmful to people, and not, as we are suggesting, on balance beneficial. By the time we weed out the emotive statements such as 'contraception is one of the curses of the twentieth century' we are left with a set of dubious factual claims. It is of course true that the divorce rate is higher than it's ever been, but in order for this to support an argument against contraception one would also have to show both that the increase in the divorce rate is, on the whole, harmful to people, and that it is caused by the availability of contraception.

Clearly we cannot decide some of these factual questions without factual evidence which is unavailable to us here. What we can do, however, is point to what would be required to make this case against contraception.

In any case there are a lot of people who undoubtedly have benefited from the availability of contraception. There are women who have already had as many children as they want or can afford, who are able to continue with their sex lives without the fear of creating an unwanted child. This has not led them to be promiscuous, and so cannot be condemned on that count, even if for some reason promiscuity were harmful. Moreover, contraception can consolidate rather than undermine a marriage.

Finally, even if it were the case that the widespread use of contraception led to the break-up of monogamous relationships in many cases, and if one believed this was harmful, there would still be an overwhelming case for the limited use of contraception as one part of solving the dire problem of overpopulation. Overpopulation causes enormous misery to countless people, and contraception can at least help to keep the birth-rate down.

The case against contraception on the grounds that on balance it harms people more than it benefits them does not seem to hold water. Because, on balance, the limited use of contraception does benefit people more than it harms them, our considerations so far favour it.

But the fact that a practice on balance benefits people is not sufficient to make it morally acceptable. It is not sufficient because sometimes one has to take into account other factors. For instance, there might be a situation where people as a whole would be better off if a certain person were dead. Imagine someone who is very wealthy, and also very mean. If he died sooner, rather than later, his son, who is benevolent, would inherit his father's fortune, and put it to a use which would benefit many people. So, to kill the wealthy man would, let us assume, produce a net benefit for mankind. Nevertheless, many people would still maintain that it was wrong to kill him, because everybody has a right to live, and you can't sacrifice someone for the general interest of other people.

But in the case of contraception we can say that nobody's rights have been infringed. The only obvious candidates would be the children who were never conceived. But it would be absurd to say that their rights have been infringed, because they never existed in the first place.

As contraception doesn't infringe anybody's rights we can argue about its moral acceptability *simply* by concentrating on the relative benefits and burdens it brings to people. We have assumed that an action is morally desirable if it produces an overall benefit for people, as long as it doesn't infringe anyone's rights. Limited use of contraceptives does, we believe, produce an overall benefit for people, and as we have just explained, it doesn't infringe anyone's rights.

Even if you are quite sure that contraception is morally desirable there is still a long list of related questions which you could usefully ask, but which we cannot go into here. Do you think that contraceptives should be available to schoolchildren? Should contraceptives be provided free on the National Health Service? Is it wrong to have sex with someone if you would not be prepared to bring up a child, were one conceived, and your method of contraception is, say, only 99 per cent effective? In order to answer these further questions it would be necessary to find out which of the alternatives, on balance, benefited people the most, and how the rights of the various people concerned would be affected by each of the alternatives.

ABORTION

All too often people haven't used contraceptives, or the contraceptives they use don't work, and the woman concerned finds herself unwillingly pregnant, and in many cases desperate. If the pregnancy is allowed to run its course an unwanted child will be born. Many people who believe that there is nothing morally abhorrent about contraception draw the line at abortion. They argue that there is a fundamental difference between the two practices, and that abortion is wrong. Killing a foetus is, they argue, taking an innocent human life, and as such is really tantamount to murder. People who defend abortion are no less opposed to murder, so if they are to defend the practice of abortion, it is necessary for them to be able to show that it isn't really murder – or at least, that if it is, it's murder with a very big difference.

Most of our discussion of contraception concerned the issue of the consequences of contraception. This is because contraception fairly clearly need not infringe anyone's rights. With abortion the main emphasis is on the question of the rights of the foetus. If the foetus is entitled to the full complement of human rights, then even if abortions have beneficial consequences for people, the morality of abortion would be highly questionable. Therefore we shall only *mention* the consequences of abortions for people, but discuss at some length the more difficult problem of whether or not the foetus does have human rights.

The arguments for the desirability of abortion are similar to those for the desirability of contraception. As far as the people who have conceived the foetus are concerned, abortion could be seen as a last resort if contraception has failed or, for some reason, not been used. The availability of abortion thus provides an extra guarantee that people's sexual encounters will not lead to the creation of an unwanted child. Individuals can have good reasons for not wanting children, and groups of individuals can have good reasons for wanting people not to have babies. Abortion can thus serve individuals and groups in much the same way as contraception can.

Are there any arguments against abortion based on assessing its consequences? First of all, it is quite clear that undergoing an abortion can be traumatic for the woman concerned. It can certainly involve much emotional stress. Furthermore, abortion puts a far greater strain on health care resources than does contraception.

But these facts are readily admitted by the defenders of abortion. They would concede that contraception is on all counts a superior method of birth control to abortion. Nevertheless, the trauma a woman might go through by having an abortion is in many cases outweighed by the benefits of not having to carry and give birth to a child whom she doesn't want, or can't afford to bring up. All that follows from the above argument is that abortion should not be promoted as an alternative to contraception. The question, however, is not whether or not abortion is a suitable alternative to contraception, but whether it should be available as a safety net when contraception has failed.

Most of the arguments about abortion centre on the rights of the parties centrally involved – the mother and the foetus. Anti-abortionists argue that the foetus has the rights of a human being, a defenceless one at that, and that having a foetus aborted amounts to having an innocent person killed. On the other hand pro-abortionists argue that the foetus, unlike its mother, is not really a person. It is technically a parasite, and should be treated as such. Just as people are entitled to have tapeworms removed from their bodies without doing wrong, so too should mothers be able to have foetuses removed.

People become passionately involved in arguments on this subject, and often make direct appeals to the emotions. You might infer that questions about the rights of foetuses and mothers are simply matters of personal feeling, and that therefore there is nothing to argue about. Certainly nothing could *prove* that foetuses have rights, or that they don't; but it doesn't follow from this that one can't have good reasons for believing that they do or that they don't. What philosophy can do here is to focus on the actual arguments that are used by both sides, see what their strengths and weaknesses are, and see what values are appealed to, and whether they are coherent.

In Britain at the moment a woman is legally permitted to have an abortion provided that two medical practitioners are of the opinion:

(*a*) that the continuance of the pregnancy would involve risk to the life of the pregnant woman, or of injury to the physical or mental health of the pregnant woman or any existing children, greater than if the pregnancy were terminated; or
(*b*) that there is a substantial risk that if the child were born it would suffer from such physical or mental abnormalities as to be seriously handicapped. (Halsbury 1969, p. 682)

The law is officially interpreted to mean that the injury to the pregnant woman or her children need not be 'grave or permanent', and that in assessing the likely effects on the health of the woman and her children, environment shall be taken into account.

This means that, in effect, no woman is barred by law from having an abortion in the early stages of pregnancy, because normally the risks of death due to a vaginal extraction abortion are substantially lower than those due to childbirth. And this aside, if two doctors believe that continuing the pregnancy is likely to damage the physical or mental health of the mother or her children only very slightly more than having a termination, she is allowed to have an abortion.

The main constraints on women having abortions in Britain are financial. With the health service over-stretched, it is often not possible to have an abortion on the health service, because of the length of the waiting list. This means that women must go to private clinics, which many cannot afford. Pro-abortionists in the Women's Movement want abortion on demand – invoking 'a woman's right to choose'. This would involve more resources being spent on health service abortion units, and the continuation of something like the present legislation.

Many opponents of abortion claim that they are not against abortion as such, but only abuses of the practice. Nevertheless, the reforms they are demanding show that they want a very severe restriction on the rights of women to have abortions. It is clear from the statements both of the interestingly named Society for the Protection of the Unborn Child, and of Life, organisations which have supported the Benyon amendment to the 1967 Abortion Act, that they are opposed to abortion not just on demand, but in nearly all cases. This is not surprising, because if the issue is really about whether or not the foetus has rights, as human beings have rights, there is a very big gap indeed between those who believe they do and those who believe they don't.

Most anti-abortionists believe that abortions are permissible only in very extreme circumstances, when abortion is the only way of saving the mother's life, or at least the only way of saving her from 'grave and permanent' injury. The fact that even anti-abortionists concede that abortion is sometimes morally permissible is borne out in the 1967 law. Though conscientious objection to performing abortions is a sufficient reason for doctors not to have to perform them, nothing in the act

> shall affect any duty to participate in treatment necessary to save the life or to prevent grave permanent injury to the physical or mental health of the pregnant woman. (Halsbury 1969, p. 685)

Some people think that the mere avoidance of grave permanent injury to the woman is not sufficient to justify killing a foetus; but very few would go so far as to deny an abortion to a woman who will die if she doesn't have one.

Allowing abortion in these very restricted cases is consistent with the belief that the foetus has the rights of a human being. One can appeal to principles of self-defence. In general it is legally permissible to kill someone in self-defence. Such killing is morally accepted by most of us, provided it is the only way of defending one's own life. There are two objections to this comparison.

First, someone might say that abortion is relevantly different from killing in self-defence because the foetus, unlike someone trying to kill you, is not a morally responsible agent, and so is innocent, not to blame, and therefore should be allowed to live, even at the expense of the mother's life. The answer to this is that the self-defence argument doesn't in its general application rely for its strength on the supposition that the would-be killer is morally responsible: perhaps he is a homicidal maniac who cannot control what he does, and therefore cannot be held morally responsible for his actions. Although he is a threat to someone's life, he is just as innocent as a young child, or a baby, or a foetus – he has no control over what he is doing. So, if it is justifiable to kill in self-defence, it is not on account of the moral responsibility of the person who is threatening one's life. If this is so, then the foetus is not in a special position because of its innocence.

The other objection to the view of abortion as self-defence is that it is not really like a standard case of self-defence because the woman is responsible for the existence of the being she is going to kill. In response to this one can say that even if the argument were accepted it would not apply to women who were not responsible for the conception of their foetuses, as in rape cases. But what about cases where a woman has deliberately conceived a foetus? If you believe that a foetus has human rights, then this case would be comparable to inviting someone on a boat trip and then, through adverse fortunes, becoming short of provisions. If only one of you can survive, many would argue that the person you have brought along should be the one to survive. After all, you are responsible for bringing him here. You got him into this mess. If the case of the foetus's survival threatening that of the mother's is like this, it really isn't strictly analogous to killing in self-defence. What is clear from all this is that if you accept that foetuses do have human rights you would want there to be substantial changes in the abortion law as it now stands.

Let us now look at arguments people bring forward on behalf of the rights of foetuses. The arguments do not try to show that adults have rights to life. This is assumed. They argue that foetuses have rights equal to those of infants, and that infants have rights equal to those of children, and that children have rights equal to those of adults. But why should foetuses have rights?

The short answer given to this is that they are human beings. The difference between a newly born baby and a foetus which is just about to be born is simply one of geography. But a simple matter of geography is not important enough to make such a moral difference. If you believe that abortion where the mother's safety is not at stake is permissible, then you ought, in order to be consistent, to believe that infanticide is in principle permissible. But most of us believe that infanticide is wrong because it is the taking of an innocent human life.

Many anti-abortionists argue that the foetus becomes a human being at the moment of conception. The moment of conception is thought to be crucially important because it is believed that after the moment of conception, but not before it, there is a unique individual with its full genetic complement.[1] The point is that before conception there were just millions of sperms and an egg. There is no individual who is the 'victim' of contraception. After conception there is at least some sort of victim which stands to gain or lose.

Another argument is that if you don't decide that conception is the relevant moment, after which a new person exists, it is difficult to know where else one could plausibly draw the line.

The basis of the anti-abortionist position is that foetuses are relevantly similar to people. Therefore they ought to be treated in the same way. But we all agree that it is wrong to kill innocent people. Therefore, treating foetuses in the same way, we ought to think it wrong to kill foetuses.

1 In fact it is possible for a fertilised egg to split to form identical twins, so strictly speaking you can be sure that there is a unique individual only after the zygote has been implanted in the womb, which happens several days after fertilisation.

How, then, would an argument in favour of abortion go? One strategy is to try to show that the argument against abortion leads to absurd consequences, which even opponents of abortion would not accept. This method of argument, called *reductio ad absurdum* ('reduction to the absurd'), is very widely used in philosophy. Here the argument might be something like this. The claim that foetuses are, from the moment of conception, human beings, would rule out as murderous what is really just a form of contraception – the intra-uterine coil or 'loop'. It would also exclude such forms of birth control as the 'morning-after pill'. The coil prevents a newly-fertilised egg from embedding itself in the uterine wall, and developing into a growing embryo. There must be something wrong, so the argument goes, with a doctrine which leads to the absurd conclusion that the disposal of a bunch of cells is murder. In the very early stages of pregnancy, it is just wildly implausible to regard the foetus as a person. Of course it is true that it is a living organism, and it is also true that it is a *human* embryo, as opposed to, say, a canine one. But it doesn't think, feel, move about, or do any of the things characteristic of human beings. There is all the difference in the world between such an agglomeration of cells and a child, or even a new-born baby. The opponents of abortion would have us ignore these striking differences. It is of course true that there is not much difference between a new-born baby and one which is about to be born. But most pro-abortionists need not, and would not, deny this.

The argument that since there is not a clear-cut distinction between the various stages in the life of a human being after conception, conception is the only reasonable cut-off point, is not convincing. This can be shown by analogy. Imagine an artist about to paint someone's picture – say it's Rembrandt about to paint his mother. He begins with a blank canvas. There is no picture at all until the first stroke of his brush. But is there a picture after the first stroke? Is there already a work of art? Surely the picture emerges slowly as a picture. If the canvas was destroyed just before the first stroke of the brush, in no sense would a picture have been destroyed. But if it was destroyed after just one stroke, would there really be much difference? Imagine the artist's son putting his foot through the canvas either just before or just after the first stroke. In both cases an appropriate attitude towards him might be one of mild annoyance. Both are a far cry from the tragedy which would result from his doing this to the almost-completed portrait.

We don't wish to argue that people are like works of art in every respect. However, there are striking similarities. To place so much weight on the point of conception is as irrational as it would be to place a lot of weight on the first brush stroke. There is all the difference in the world between a completed picture and one which has only just been started. Similarly there is all the difference in the world between a newly-fertilised ovum and a fully-formed person.

A pro-abortionist could argue that the more a foetus becomes like a fully-formed person the graver it becomes to countenance an abortion. So the later on a woman is in her pregnancy, the stronger the justification that would be required for an abortion to be warranted. Killing a newly-fertilised ovum is a far cry from infanticide, whereas killing a foetus which is just about to be born is as near as makes no difference to infanticide. It's all a matter of degree.

There are many people who think that there is a rational cut-off point between conception and birth, beyond which termination is a form of infanticide. This is the point at which the foetus becomes 'viable', in other words the point beyond which it could survive independently of its mother. Before this time the foetus is simply a parasite in the mother's body, but afterwards, since it could exist on its own, it is a separate individual with human rights. This argument is reflected in Britain in the Infant Life (Preservation) Act of 1929, which still stands, apart from minor terminological changes. The Act states that

> Any person who, with intent to destroy the life of a child capable of being born alive, by any wilful act causes a child to die before it has an existence independent of its mother, shall be guilty of felony, to wit, of child destruction, and shall be liable on conviction thereof, on indictment to penal servitude for life . . . For the purposes of this Act,

evidence that a woman had at any time been pregnant for a period of twenty-eight weeks or more shall be prima facie proof that she was at that time pregnant of a child capable of being born alive. (Halsbury 1969, p. 304 f.)

It is interesting that the Act also exonerates anyone who 'destroys the life of a child capable of being born alive' if this action is necessary to save the the mother's life.

As the law stands now in England, then, if a pregnancy is terminated after twenty-eight weeks the person who terminates it is guilty, in effect, of homicide. The law is similar in the United States. Recently an American surgeon from Boston was convicted of manslaughter on the grounds that he had terminated the pregnancy of a woman whose foetus was viable, and had not attempted to save its life after the termination. During the trial the prosecution showed the jury photographs of foetuses in their late stages of development. The prosecution described them as unborn babies. Despite the judge's directing the jury not to be swayed emotionally by these photographs, it is difficult to ascertain whether their verdict of guilty was reached emotionally or whether they were moved by rational considerations.

What should we say about viability as a criterion? First of all we must be firmly on our guard against surreptitious question-begging. The Infant Life (Preservation) Act and many opponents of abortion preempt discussion of whether it is wrong to kill foetuses after they have attained viability by simply defining a viable foetus as a 'child'. Most of us are against killing children, and so if these foetuses *are* children, we ought, out of consistency, to be opposed to killing them. But simply calling them children doesn't make them children. Thus we should be wary of the group which calls itself the 'Society for the Protection of the Unborn Child'. The whole issue is about whether or not foetuses are children at all.

So, leaving all these emotional tactics to one side, what can be said about the argument for making viability the crucial cut-off point? With the availability of effective incubators foetuses attain viability somewhere between the twenty-first and twenty-sixth weeks of a pregnancy. At this stage foetuses are pretty well formed and have striking resemblances to newly-born babies. But the stage at which a foetus can exist independently of its mother is a function of technological resources and know-how. James White's Select Committee on abortion law reform proposed that the 1929 Act be changed so that it would be a serious offence to terminate a pregnancy after twenty weeks. It might well be the case that in ten years or so it is technologically possible to develop fertilised human ova into babies outside a woman's body. It might become possible to transfer a newly fertilised ovum from one woman to another, thus making it not totally dependent on its own mother for survival. This is already happening with sheep. If these advances occur, then viability might come to coincide with fertilisation, and so the crucial date will have been brought forward to where the strictest Roman Catholic would put it.

Thus the main objection to viability, or being 'capable of being born alive', as a criterion for deciding when it is and when it is not morally permissible to terminate a pregnancy, is that viability is determined by the skills and resources of a particular society, and ignores the state of the foetus itself. What determines whether or not it is all right to kill the foetus is whether or not it is a person, and it is just not plausible to claim that the stage at which a foetus becomes a person alters according to the technological competence of a society. Suppose, for example, that we had developed a special kind of incubator which could enable a foetus to develop outside its mother's womb from twelve weeks onwards. If viability were the criterion of being a person, then from this time on all twelve week-old foetuses in their mothers' wombs would suddenly have started being people. You would also have to accept the strange conclusion that if the machines all broke down these foetuses would suddenly stop being people – at least until the machines were repaired or replaced. Whether or not something is a person cannot plausibly be decided in such a way.

Is a foetus a person? This seems to be the crucial question, because most of us believe that people are special. They have a number of rights, including the right not to be killed for the convenience of others. In deciding the issue of whether or not a foetus is a person, one appears to face a dilemma. Either one accepts that foetuses are people, in which case using the 'loop' or the 'morning after pill' is as bad as murder; or foetuses are not people, in which case there is nothing wrong with killing a nearly-born foetus, even though it is not significantly different from a newly-born baby.

It is easy to be misled by such questions into attaching too much importance to finding a clear-cut answer to this question. It is essential that one be on one's guard against being led up blind alleys by one's questions. In this context the word 'person' often leads people to pursue bogus proofs. Whether or not a foetus of a given age is a person is of course important, but it doesn't follow from this that there is a clear-cut answer to the question 'When does a foetus become a person?' To believe that this question has a precise answer is like believing that there is a definite answer to the question 'When does a series of brush strokes become a picture?' Notwithstanding this, we can say that what is on one canvas constitutes more of a picture than what is on another. As we said earlier, pictures emerge gradually. So do people.

A newly-fertilised ovum has very few of the properties of a person. It doesn't breathe, it doesn't move, it doesn't have any limbs or organs, and it lacks a brain. On the other hand a nearly-born foetus does have its full complement of organs. It has skin, a brain, and a heart which beats; it moves, and as soon as it is out of the womb it will cry and start breathing. No matter where you choose, if you choose, to draw the line, it is clear that the almost-born foetus has many more of the properties of a person than the newly-fertilised ovum. Nonetheless, the nearly-born foetus lacks a number of the properties of fully-fledged human beings. In particular it lacks the higher intellectual skills which go a long way to differentiating us from other animals. A child becomes a fully-fledged person only after several years of life.

In trying to decide about abortion, rather than ask the simple-sounding question 'Is a foetus a person?', it would be more profitable to ask at various stages of its life something like 'How marked are the similarities and differences between this foetus now, and a complete person?' Having answered such a question you can then ask yourself whether you think these differences are sufficiently great to warrant treating the foetus at the particular stage of its development quite differently from a fully-fledged person. Such an approach avoids the implausible assumption that there is some magical moment when all of a sudden a foetus is transformed into a person. Becoming a person is a gradual process. By the same token being a person is a matter of degree. Young foetuses are less persons than older foetuses, who in turn are less persons than young babies, and so on until a child becomes a fully-fledged person.

If being a person is a matter of degree, then one can plausibly argue that the acquisition of rights is a matter of degree. A very young foetus, being hardly a person at all, has at most minimal rights. On the other hand, a twenty-eight-week-old foetus is much more of a person, and has more rights. A sliding scale such as this reflects many of our basic intuitions, for example that there is a greater need to justify a late abortion than an early one, and that infanticide is worse than abortion, but not as bad as ordinary murder. Of course, even if you accept this view, the problem remains of deciding exactly what you think are the crucial stages of development in an embryo/foetus/baby/child. Biology and developmental psychology should help here.

If our view of the development of a person is correct, then, discounting side-effects, abortion on demand is morally permissible in the early stages of pregnancy. As the foetus develops it would become increasingly difficult to decide. Perhaps after about twenty-four weeks the foetus is sufficiently developed to have rights which would outweigh trivial reasons for abortion, such as the desire not to be pregnant over Christmas because then you'd have to postpone your skiing holiday. One would be deciding to take a very grave step – to kill a creature which was very like a young person. In addition, the costs to the mother and to the health services would be that much greater. But these side-effect arguments, though very important, are not part of our main discussion here, which has been about whether we are justified in treating foetuses with less than the respect we give to mature human beings.

Summary of the moral argument Let us summarise the moral argument put forward here. It is not in principle wrong for women to have abortions. In order to make a judgement about a particular case one has to take into account all the parties concerned. We have argued that the foetus in its early stages of development is so far removed from being a person that it does not have the rights to life which most of us believe are due to human beings. It gradually acquires them along with human characteristics. To lay down fixed rules about exactly what is morally permissible is not feasible if one accepts the view that becoming a person is a continuous process rather than a sudden step.

Abortion and the law Even if there are no clear-cut guidelines which tell one under what circumstances termination is morally permissible, this does not imply that there should not be clear-cut legislation. The consequences of having ill-defined laws are often harmful, because the law can be misinterpreted: so there is a case for stipulating what is and what is not legally permissible, in precise terms, even if one's stipulation does not perfectly mirror what one believes is intrinsically right or wrong.

In drafting laws legislators must not ignore any of the foreseen consequences of the legislation. One of the main slogans of the National Abortion Campaign is 'No return to backstreet abortions'. Before abortions became easier to obtain after the 1967 Act, many women were forced to have back-street abortions because they were so desperate not to have a child or, in many cases, yet another child. The existence of legislation forbidding abortions did not mean abortions didn't occur at all. What the law did mean was that abortions took place in highly unsanitary conditions, often resulting in infection and even loss of life for the unfortunate women who resorted to the illegal practice. One can draw an interesting parallel here with the anti-liquor legislation in the United States. One might think that drinking spirits is wrong because of the harm it does to people. But making it illegal to drink whisky does not mean that whisky will not be drunk at all. Instead, the whisky people buy – on the black market – is likely to be of inferior quality, containing dangerous impurities. During the prohibition on the sale of liquor in the U.S.A. many people were poisoned by the impurities in the alcohol they bought.

But the two cases are in fact not strictly analogous. The opponents of early abortion believe that even embryos are people. So they believe that every time a woman has an abortion a person has been murdered. If you believe this you might still argue for strong anti-abortion legislation even if this led to the deaths of a number of women, because in all likelihood their number would be substantially lower than that of the foetuses who would be saved from abortion. If anti-abortion legislation would reduce the number of abortions carried out, whether in the back streets or elsewhere, by more than the number of women who would die as a result of illegal abortions, then if you believe that it is wrong to kill people, and that foetuses really are people, you ought to oppose liberal abortion laws.

This consideration forces us to focus our attention once again on the central question of whether or not we are to regard foetuses, at various stages of development, as people.

INFANTICIDE

People who are opposed to abortion often defend their position by claiming that abortion is really no different from infanticide. What should we say about this move?

First of all, given the continuum account which we have been putting forward, it is quite true that there is not a significant difference between killing a developed foetus and killing an infant. As we have said, the difference between a nearly-born foetus and a newly-born baby is essentially one of geography, and where human rights are concerned, such a geographical difference hardly seems significant. In other contexts we would not claim that a person should lose his human rights simply because of his geographical location. But it does not follow from this that there is no important difference between killing an infant and killing a newly-conceived foetus. As a foetus develops it acquires many of the properties which entitle people to rights. And clearly an infant, which is still more of a person than a developed foetus, has still more of the relevant human characteristics, and so is entitled to more of our respect. The older an infant becomes the more like a person it grows, and so the more serious it becomes to kill it. According to our view it is morally justifiable to kill a young foetus for a relatively trivial reason, but it would certainly not be justifiable to kill an infant, except for a very important reason.

There are two sorts of reason for killing an infant, just as there are for having an abortion. The first centres on the interests of people other than the infant; the other on the interests of the infant. Of course the two might well be connected. As infants are so nearly fully-fledged people, reasons of the first type would have to be pretty well a matter of life and death. For instance, it might be justifiable to kill one's infant if this was the only way of ensuring that one's other children did not die of starvation.

The case for infanticide in Britain is strongest when reasons of the second kind apply. Quite often babies are born with mental or physical handicaps which are so severe that they are likely to have miserable lives if they are allowed to go on living. Many think it would be a kindness to kill them, that they would be better off, so to speak, dead than alive. The motivation behind this is like that behind euthanasia. The main difference is that the infant does not have the capacity to make a judgement on the issue. Is it right to take such a momentous decision on behalf of the infant? It is difficult to see anything wrong with it in principle, because nobody's interests are being trampled on. The great difficulty comes with assessing whether it is in the infant's interests to be killed. There are cases of people with the most severe handicaps living worthwhile and relatively happy lives. Furthermore, parents and close relatives might have a strong reason for wanting their child to die, and this may cloud their judgement. This means that even if infanticide became permissible one should still give it extremely careful thought.

One should also not ignore the side-effects of one's actions. If you do kill an infant it might either cause you great unhappiness through feelings of guilt, or you might lose some of your qualms about killing in general, and end up killing people unjustifiably; or your actions might influence others to do this. It is also argued that bringing up a handicapped child can have a very beneficial effect on the personalities of the parents. An individual would still have to face these problems even if infanticide were legal. But should it be legal?

Infanticide and the law We have already argued that what one thinks the law should permit and what one holds to be morally justifiable don't necessarily entirely correspond. In the case of abortion, for example, there is at least a prima facie case for permitting abortions, even if you think abortions are wrong. With infanticide it seems to be the other way round. There's a case for not permitting infanticide even if you think it's sometimes morally justifiable. The enormous difficulties of producing a law which permits justifiable infanticide while excluding other cases are so severe that it is better to have relatively tight laws restricting the practice, even if you think the practice isn't wrong in itself.

Probably the most important difficulty which confronts legislators drafting a law which would permit infanticide in some cases, is determining the criteria which would be applied. First, it would be extremely difficult to draw up legislation which would ensure both that the rights of the infant are not overridden for the wrong reasons and that an infant can be killed if its continued existence is going to threaten the lives of existing people. Secondly, it would be even more difficult to draw up criteria which would determine whether an infant's life would be so bad as to be worth not living.

Although one might argue, as we have done, that there is a variety of circumstances in which one can kill an infant without committing an injustice, it is very unlikely that laws could be drafted which permitted infanticide in just these cases, without sometimes allowing infants to be unjustly killed. Thus if we believe that the law should provide strong protection for people against unjustified killing, we might oppose legislation which permits infanticide, even if infanticide is sometimes morally justifiable.

The other argument against legalised infanticide has already been mentioned in the previous section. If infanticide were made legal, perhaps sooner or later people would come to value life less than they do now, and this would lead to an increase in unjustified killing. In fact there is a number of peoples who practice infanticide, but nonetheless place a high value on the rest of human life. Whether or not the legalisation of infanticide in this country, now, would diminish people's respect for the rest of human life, is a question which, if it can be settled at all, can be settled by social scientists. In any case, it is not one we need to rely on.

Much more could be said on the issue of whether infanticide is ever morally justifiable and whether it should, in some circumstances, be legally permissible, but space and the topic of the book prevent a more detailed treatment. Also there is a number of important further questions, such as 'How severe should the penalties for infanticide be?', which we can't discuss here. But we do hope to have said enough to show what kinds of consideration philosophy uses as the basis for moral conclusions in this area.

EUTHANASIA

As we began by saying, one of the more dramatic consequences of recent advances in medicine is that we can now prolong the lives of people who would otherwise have died. We now have controversial cases such as that of Karen Quinlan, an American girl who went into an irreversible coma after attending a party with her friends. She was kept alive by a mechanical respirator and a wide range of antibiotics. She was in a 'vegetative state' from which, barring a miracle, there was no chance of recovery. After a lengthy legal battle her parents eventually won the right to have the prolongation of her 'life' discontinued. Increasing numbers of people believe that this is a precedent to be welcomed. In a recent speech, for instance, Dr Coggan, the Archbishop of Canterbury, said that

It was wrong to prolong life 'just for the sake of doing so', and a fallacy to think that the Christian view was in favour of life at any cost. (*The Times*, 14 December 1976, p. 1)

He contrasted the mere failure to prolong someone's life with euthanasia, which involves the deliberate shortening of someone's life. The question of what, if anything, are the significant differences between euthanasia, and the mere failure to keep alive at all costs, is important and perplexing, but we shan't be discussing this issue here. We shall be talking specifically about euthanasia.

The morality of euthanasia Suppose that someone has developed an extremely painful terminal illness, and prefers to die sooner rather than prolong the agony. Should it be possible to alleviate this suffering by giving him a drug which will painlessly kill him? To do this is illegal in the United Kingdom, and would be frowned upon as immoral by many. It was even possible to be sent to prison for attempted suicide in Britain until quite recently. The law, and attitudes, have changed with respect to suicide, but euthanasia still remains illegal, and frowned upon by many. What, if anything, is wrong with euthanasia?

Killing, or helping to commit suicide, someone who wanted to die with good reason, and asked you to help him, doesn't seem to infringe anyone's rights. Human rights include the right to do what you want, at least provided that you don't infringe other people's rights to do what they want. It is on these grounds that many regard suicide as a human right. But euthanasia is in many respects no different from helping someone to commit suicide. This is so in cases where the person would commit suicide, but lacks the wherewithal.

Euthanasia is sometimes called 'mercy killing'. This is because people are sometimes so badly off that it is kind to them to 'put them out of their misery'. Clearly euthanasia can relieve much suffering – both for the person wanting it, and for those near and dear to him.

So, barring any specifically religious argument, such as the claim that euthanasia is wrong simply because God wills it that way, the practice, at least where the person requesting it is of sound mind, does not pose any special moral problem. The moral difficulties crop up with people who are not of sound mind. Many good candidates for euthanasia are either 'vegetative' or senile, and thus not able to make a rational choice. In both cases one is faced with what might be called the problem of paternalism. Is it right to take such a momentous decision on another's behalf?

As far as we know, the person in a vegetative state does not have the conscious experiences which make life worthwhile. Such a person would not lose anything by being killed, unless of course there is a chance of recovery of consciousness. One could bring about the shortening of this person's life either by not administering drugs which are required, or by the positive action of switching off a respirator or some such. We are discussing the latter alternative. By switching off the respirator much suffering can be avoided. For example, some of the scarce resources spent on keeping the person alive could be diverted to other people who would otherwise be suffering for lack of medical care. Futhermore, the person's friends and family could be spared considerable pain. The problem here is 'How small must be the chances of recovery before it is permissible to kill a human vegetable?'

The case of senile people is more problematic. One's motives might well be suspect. Suppose granny is getting old, has developed bad aches and pains, and is miserable because granddad and many of her friends have died: her future doesn't look at all bright. Suppose also that her children, who look after her, find her a terrible nuisance and worry, and that when she dies they will inherit a lot of money. They might argue as follows. 'She is really very old, and doesn't enjoy life. In fact life is a misery for her. She would really be better off dead now, because if she lives on she'll only get worse and will become increasingly miserable. If she could only look at things rationally, which unfortunately she can't because she's senile, and anyway we're all afraid of death, she would agree that it was in her best interests to be dead. Let's see if we can persuade her that this is so.' The result of all this might be that their willingness to see her dead kills her will to survive, and is crucial in persuading her to sign her own death warrant. The relatives might consciously think they are acting in her best interests, but nonetheless there does seem to be something abhorrent in this kind of behaviour.

There is a clear difference between killing someone because he wants to be killed, for his own sake, and killing someone for one's own convenience. In cases of euthanasia where a senile relative is involved there is always a chance that one might be putting on the pressure for selfish reasons.

Another difficult case is someone who is going through a temporary depression, on the basis of which he wants to be dead. Here the problem is that if he somehow manages to get through this temporary crisis, he will be able to live a worthwhile life, and will be thankful that he is still alive. Most people would think that under these circumstances it is right to act in a paternalistic way, to protect the person. But there are grave difficulties in determining whether a depression is likely to be permanent.

Because of all these essentially practical difficulties of determining one's motives, and whether it would be better for the person to die sooner rather than later, extreme care should be taken before being a party to euthanasia, which is, after all, irreversible.

Euthanasia and the law As euthanasia can eliminate much suffering, then if no one is harmed by it, there is a good case for making it permissible, at least under some circumstances. However, because of the grave temptations to abuse the practice, say by putting pressure on people to sign a form consenting to euthanasia, we think there should be fairly strong legal checks.

Many of the requests for euthanasia would come from people in the later stages of a painful killer-disease like some types of cancer. In such cases doctors would obviously be relevant people to consult, because they more than any other people would be able to give an accurate prognosis. The sort of legislation we recommend would allow euthanasia in these cases, provided, say, a doctor, or possibly two, agreed that the disease was terminal, and provided the patient and the next of kin requested it in writing. Of course, under any system there is a chance that the patient would be pressured into signing – but at least in this sort of case there would be a good reason for him to want to die sooner, and the suffering which could be avoided is enormous.

Euthanasia should, we think, be available for people in permanent vegetative states. Here, the main motivations for killing are to be able to use scarce resources where they can be more effective, and to save friends and relatives from suffering. As long as the person would anyway never recover from the condition one doesn't harm him by switching off the respirator. So we think the law should allow this practice because it benefits some people and doesn't harm anyone. Again, doctors' opinions should be sought because they are the people who are trained to know the chances of recovery for someone with a damaged brain. There is a good reason for requiring that the person's next of kin, assuming he can be found, and is of sound mind, should agree, because this decision is likely to affect him more immediately than other people. Also, having an opinion other than the doctors' is a check against doctors being over-pessimistic in their prognostication; not that this would be a widespread phenomenon anyway.

This leaves the most difficult case, euthanasia for someone who is just depressed, wants to die, but is not brave enough to kill himself. This would include both young and old people.

There are two major problems with permissive legislation in this area. The first is that it might lead people who are temporarily distressed to seek euthanasia when, if they lived, they would recover from their distress and live a worthwhile life. An example of this might be someone who had just lost a loved one.

He might feel absolutely desperate and suicidal for a limited time. If it was possible to have euthanasia on request, many such people might be killed for insufficient reasons.

The second major difficulty stems from the real possibility of abuse. Even as things are now, many old people, submitting to pressure from relatives, sign forms agreeing to committal to an old people's home, even if they don't really want to go to one. If euthanasia on request were available, many, especially old people, might end up signing away their lives, at the suggestion of other people with vested interests in their death. There would actually be an incentive for relatives who for selfish reasons wanted a person out of the way, to make his life more miserable than it need be.

These difficulties with avoiding misuse or abuse lead us to oppose permissive legislation for the kind of euthanasia that is justified only on these psychological grounds.

What is more important here than our actual conclusions are the reasons for them. We do not oppose voluntary euthanasia on principle, and in fact are in favour of its limited use, with fairly strict checks, in the case of people in permanent vegetative states, and people who seek it, if they have painful terminal illnesses. Our reasons are based on the benefits which euthanasia can bring to people.

COMPULSORY TREATMENT OF MENTAL ILLNESS

According to the census of patients in mental hospitals and units in England and Wales which took place on 31 December 1971, 7,177 people were legally detained. This means that they were in an institution whether or not they wished to be. The treatment of mentally ill people raises many of the most profound issues in social philosophy – problems of liberty, human rights, and welfare. In the next few pages we shall look briefly at the law in England and Wales (the law is slightly different in Scotland), see where the law is ethically objectionable, and propose directions in which we think it should change.

John Stuart Mill wrote over a hundred years ago, in his celebrated work *On Liberty*:

> . . . the sole end for which mankind are warranted, individually or collectively, in interfering with the liberty of action of any of their number, is self-protection . . . the only purpose for which power can be rightfully exercised over any member of a civilised community, against his will, is to prevent harm to others. His own good, either physical or moral, is not a sufficient warrant. He cannot rightfully be compelled to do or forbear because it will be better for him to do so, because it will make him happier, because, in the opinion of others, to do so would be wise, or even right. These are good reasons for remonstrating with him, or reasoning with him, or persuading him, or entreating him, but not for compelling him, or visiting him with any evil in case he do otherwise. (Mill 1859, p. 72 f.)

This principle of non-interference with people's lives, as long as they are not harming others, seems reasonable to most of us. Although we consider it to be the responsibility of the state to look after people's interests in order to increase their general level of happiness, most of us think there should be quite severe limitations on the powers of the state to achieve these ends.

Freedom to run our own lives is something which most of us value very deeply. In many circumstances one would put freedom above pleasure. To illustrate the truth of this, try to imagine a world where, with the aid of drugs, everybody's thoughts and feelings were controlled by a benevolent despot. People would not, in such a world, be troubled by the neuroses which bedevil most of us. They would not feel depressed, jealous, resentful, or suffer from any other painful emotions.

Most people would prefer to live in society as we know it, provided their lives are not too horrible. They would prefer life as it is, in spite of the suffering involved. Why is our world more attractive?

One possible answer is that our society is at least interesting. We would get terribly bored in a society where everything always worked out. But this answer really misses the point. For we could imagine that under a benevolent despot, the drugs were so powerful that one was never bored, but constantly stimulated.

The real answer, it seems to us, is that in the imaginary society, people would not be free, and freedom is, at least in most circumstances, more important than happiness. Even if this society would give us the illusion of freedom, few of us would opt to join it.

There are many examples from everyday life of how the state does not interfere with our freedom to act against our self-interest. For instance, many thousands of people die each year from illnesses associated with smoking. Moreover, many smokers desperately regret their smoking habits. Much misery is caused by smoking – and yet people are not banned from smoking. They are certainly not locked up in mental hospitals because they smoke. To prevent people from smoking would be a gross infringement of their individual liberty. Dangerous sports are similar. Every year people are killed in motor-racing and climbing accidents – and yet it is almost unthinkable to ban these pursuits. Remarks like 'Well, if they're crazy enough to want to do it, let them; it's their own funeral' are common. The same applies to smoking, and eating high-cholesterol foods, but, interestingly, not to using marijuana.

So, in many areas of our lives the state respects the sentiment expressed in Mill's treatise on liberty. Nevertheless, there are two categories of people whom the law protects from the harm they might do to themselves, namely, children and the mentally ill. The protection of children by laws such as that which prevents them working in factories poses extremely interesting and perplexing problems, which we shall not be able to discuss here. However, many of the considerations which apply to our treatment of the mentally ill also apply to our treatment of children.

The most important law concerning mental health in England and Wales at the moment is the Mental Health Act of 1959. The Act was seen as a great advance in the treatment and care of mental illness, because it sought to remove the stigma attached to sufferers from mental problems, and to treat them with the respect due to any fellow human being. Prior to the Act mental patients were often regarded as moral degenerates whom it was quite right to punish.

The 1959 Act reflected the optimistic hope that mental disorders were like physical ones, and that it would only be a matter of time before the medical profession discovered a cure for them. Formerly, mental patients lost their citizens' rights and were severely stigmatised. By categorising mental abnormality as an illness, some progress has been made towards removing these stigmas. A person can now go to a mental hospital, as a voluntary or 'informal' patient, and like any patient with a physical ailment, can discharge himself when he wants to. This part of the act is clearly in line with Mill's sentiment. However,

> The architects of the Act at the same time acknowledged that some forms of mental illness could not be regarded in quite the same light as a broken leg, or a case of diabetes or chronic bronchitis. The illness could manifest itself in the form of socially unacceptable behaviour, unacceptable because it presented a danger to others or to the patient. The illness might also be such that the patient would be unable or unwilling to accept the diagnosis, and therefore unable to agree to the restraint required for his own safety or the safety of others. (Gostin 1975, p. 5)

So the Mental Health Act 1959 includes provisions for the compulsory admission for 'observation' or 'treatment' of a person said to be suffering from a mental disorder which warrants his detention in a hospital, when 'it is necessary for the patient's health or safety, or for the protection of other persons, that he should be detained' (Gostin 1975, p. 23). Thus at the end of 1971 there were over 7,000 people in mental institutions who had no freedom of choice to leave or refuse treatment.

According to the law 'mental disorder' covers a very wide range of conditions, ranging from 'severe subnormality' to 'psychopathy' and 'mental illness'. Severe subnormality and psychopathy seem to be the least problematic of these categories for the lawmaker.

'Psychopathic disorder' means 'a persistent disorder or disability of mind which results in abnormally aggressive or seriously irresponsible conduct on the part of the patient, and which requires or is susceptible to medical treatment'. 'Severe subnormality' is 'a state of arrested or incomplete development of mind which includes subnormality of intelligence and is of such a nature or degree that the patient is incapable of living an independent life or of guarding himself against serious exploitation or will be so when of age to do so'.

The term 'mental illness' is not defined in the Mental Health Act. Whether or not there is a satisfactory definition is a matter of much controversy at the moment. In any case, there are certain symptoms on the basis of which a person is said to be mentally ill. These include hearing voices, or experiencing, with some frequency, any other type of hallucination; having unreasonable beliefs – for instance, that you are Napoleon, or that you can fly; having unreasonable emotions – for instance, terror at the sight of a sparrow, overwhelming grief at the death of a slug; and having inappropriate desires and preferences, such as the preference to die of starvation rather than spend a penny more than £5 per week on food.

One of the most fundamental principles of justice is that it is wrong for judicial authorities to treat people differently unless they can demonstrate that there is a relevant difference between them. The onus of proof is on the judges to show that there is a difference, rather than on the judged to show there is not.

Why should the mentally ill be treated compulsorily, while heavy smokers are left to their own devices? In order to justify the law as it stands it is incumbent on the law-makers to be able to show that there is a relevant difference between these categories of people, which justifies their different treatment.

Probably the main challenge to the whole enterprise of compulsory treatment comes from the so-called 'anti-psychiatry' lobby. This includes writers such as R. D. Laing and Thomas Szasz. They claim that the notion of mental illness is at least badly defined, and used to incarcerate people under the guise of helping them, and Szasz in particular claims that so-called mental illness is not really an illness at all. In his *The Manufacture of Madness* he claims that mental illness is no more real than witchcraft.

For our present purposes we fortunately do not have to decide whether or not mental illness really is an illness. Mental illnesses clearly have some things in common with ordinary physical illnesses, and they differ in some ways. What is important here is to decide whether or not people who are detained under the Mental Health Act constitute a special category justifying special treatment.

Laing and Szasz, then, claim that the whole notion of insanity, as it is used by the authorities, is confused and arbitrary. Rather than reflecting a real condition, it reflects some of the prevailing values of society. Thus Laing writes:

> In the context of our present pervasive madness that we call normality, sanity, freedom, all our frames of reference are ambiguous and equivocal . . . A man who prefers to be dead rather than Red is normal. A man who says he has lost his soul is mad. A man who says that men are machines may be a great scientist. A man who says he *is* a machine is 'depersonalised' in psychological jargon. A man who says that Negroes are an inferior race may be widely respected. A man who says his whiteness is a form of cancer is certifiable. (Laing 1959, p. 11 f.)

Nevertheless, whether or not there are reasonable criteria of what is mad and what is sane, there are many people who feel severe psychological distress which they would like to be rid of. Various forms of treatment, including electro-convulsive therapy, can sometimes relieve this distress. These people can be treated as 'voluntary' patients. In these voluntary cases of treatment there is no serious moral dilemma, because the patient can refuse the treatment if he thinks it isn't in his best interests (this isn't true of *all* voluntary patients, because some of them may agree to be 'voluntary' patients only after they have been threatened with compulsory treatment if they refuse voluntary admission). The cases which do present moral problems are those where compulsory admission is the only way of ensuring that a patient is treated. But what justification, if any, is there for the compulsory hospitalisation and treatment of mentally disordered people? How do they differ from 'normal' citizens?

As we have seen, there are two basic reasons given for the compulsory admission and treatment of people under the 1959 Act. These are the protection of people in society from the actions of the mentally disturbed person, and the protection of the person himself from his own actions. Let us look at these in turn, and compare each with apparently similar cases involving 'normal' people.

The threat to others Among those who are detained under the Mental Health Act for the protection of others are included both those who have been convicted of serious crimes and those who, it is feared, are likely to commit a serious crime for the first time if they are not committed.

People who are *not* judged to be insane who commit serious crimes and are found guilty by a jury are sentenced to a fixed term in jail. When that time is up, or before if they have behaved themselves, they are set free. As long as they do not commit any crimes while they are in prison they will be released when they have served their term.

The central rationale for this practice, rather than having an indeterminate sentence which can be indefinitely extended if it is thought likely that the criminal is not reformed, is a deep-seated belief that punishment should be meted out only to people guilty of crimes, and should not be more severe than is justified by the severity of the crime. In committing a crime you supposedly incur a debt to society. In serving a term in jail, or paying a fine, the debt is paid, and subsequently society has no further right to detain you.

If someone has committed a crime, but is also adjudged to be criminally insane, the situation is different. Someone who is committed to a psychiatric hospital such as Broadmoor can be detailed for an indefinite period – at Her Majesty's 'pleasure'. Why is this?

There are two basic reasons. First of all, the offender is regarded as 'sick', not as a guilty person. He is being 'treated' in the prison hospital, rather than being punished. If he is being treated, it might seem to make sense to stop the treatment only when he is cured.

But this argument is insufficient, because in other cases of treatment for an illness a patient is legally free to go, even if doctors don't recommend it. If a patient wishes to leave a hospital when he is not fit to do so, that's his own affair. The exception to this would be someone who was suffering from a dangerous contagious disease such as Lassa Fever. Here, a patient can be detained justifiably, we think, because if he is released it is very likely that other people will be infected, and thus suffer greatly, and probably even die.

This immediately brings us to the second reason for committing 'criminally insane' people for an indefinite period. If they are set free, people in the community will be at risk. If the criminal is still insane he is likely to strike again, and people must be protected. Here, the case is analogous to the Lassa Fever patient who doesn't want to be kept in hospital. He is not blamed for having the disease, but must be confined because so much is at stake, even if this involves infringing his personal rights to liberty. It is, so the argument goes, just like this with the criminally insane person.

There are two basic difficulties with this argument for the indefinite detention of mentally disturbed offenders. First of all, studies such as the Butler Report on mentally abnormal offenders (1975) include evidence which suggests that 'mentally ill' violent offenders are no more likely to reoffend than are 'mentally normal' offenders. Yet it is not permissible to detain 'mentally normal' offenders on the grounds that they are potentially dangerous.

The other main difficulty with accepting this rationale is that it could greatly extend the legal powers of the state authorities. There would be no reason in principle to prevent the authorities from extending their powers of detention to people who had not committed offences at all; both 'mentally ill' and 'normal' ones alike. If potential dangerousness was a sufficient condition for locking people up, we might come to see the incarceration of all sorts of people who have committed no crimes, but, say, come from backgrounds which make it statistically likely that at some stage in their lives they will commit a serious crime. This would certainly violate the cherished principle alluded to earlier, that it is wrong to punish an innocent person – even if it would benefit people in society.

So we reject the indeterminate sentencing of 'mentally ill'

offenders. Until the assumption that mentally ill offenders will commit crimes again is established, the sentencing isn't justified in terms of the protection of members of society. If the assumption were proved to be true, of course, there might be a case for regarding a mentally disturbed criminal in the same light as someone with Lassa Fever.

The second group of people who are detained under the Mental Health Act for the protection of others have not committed offences, but are deemed likely to do so. Is it right to compel *them* to spend time in mental institutions?

We have just argued that it is wrong to discriminate against mentally disturbed people by giving them indefinite sentences of detention. The arguments there apply equally here. A mentally disturbed person who might commit an offence is, as far as the safety of the community is concerned, no better or worse than anyone else who might commit a crime. A mentally disturbed person who might commit an offence is similar to the offenders who are detained at Her Majesty's pleasure – except that he hasn't even done anything wrong. So if it is unjust to lock up the offenders indefinitely, it is surely wrong to do so with merely potential offenders who are mentally disturbed.

These conclusions don't seem quite satisfactory, however. We have concentrated on the rights of the mentally disturbed person. But what about the rights of ordinary citizens? What about the parents of the child who may be attacked by a psychopath? What about the rights of the child? Here, similar considerations apply both to offenders who might offend again, and to potential offenders, hitherto innocent. Much publicity has been given to cases such as that of Graham Young, the convicted poisoner who was released from Broadmoor on medical recommendations, and then committed more acts of poisoning on his release. Because he was released, innocent people lost their lives. Let us try to disentangle the different threads of this debate to see what should determine our decision on these matters. What are the crucial issues?

On the one hand we must consider the rights and interests of the mentally disturbed person, whom we shall call the patient;

on the other, those of the members of the public. A law which allows mentally disturbed people to be detained indefinitely discriminates unfairly against them. Not having a law which enables the state to detain mentally disturbed people who threaten violence puts the public at an unacceptable risk. Is it possible to do justice to both parties?

Where there is a genuine conflict of rights and interests a just solution involves a compromise, which will minimise the injustice done to either party. In the present case, what is needed is some way of protecting the public from dangerous people without withdrawing liberties from innocent people. At the moment there is a distinction in the law between emergency admissions to mental institutions and long-term ones. In an emergency, for example where someone with a knife is threatening to kill the first redhead he sees, it is possible to hold the person in custody – either at a mental institution or at a police station – for 72 hours. If the dangerous person is in a public place he can be taken to a 'place of safety' by a police officer. If he is not in a public place, emergency admission requires the recommendation of one medical practitioner. To detain a patient for more than 72 hours requires a more rigorous procedure.

This distinction between emergency admissions and long-term ones makes good sense, because swift action may be genuinely required to prevent, say, a murder being committed, and there simply isn't time to acquire the doctor's signatures. Although the dangerous person does suffer indignity and trauma, at least the detention cannot be extended indefinitely without recourse to more stringent processes.

Long-term detention is a far more serious matter for the patient, and consequently there is a case for many more checks to preclude people being detained for long periods. In Britain now there are strict safeguards to protect the rights of patients who might be locked up indefinitely simply because they are thought to be a danger to society.

Nevertheless, people who are wrongfully arrested by the police are able to sue for damages if they can prove they have been wrongfully detained. If we are really to treat 'mentally ill' people with the same respect we show for others, there should be a similar provision for them, so that they can claim damages in respect of wrongful detention under the Mental Health Act.

Our case has so far been based on the claim that there is no relevant difference between the 'mentally ill' and 'mentally normal' which justifies the compulsory detention of only the 'mentally ill' on the grounds that they pose a threat to others. Unless a relevant difference can be shown it would seem to be arbitrary and unjust to treat mentally disturbed people as a special category. If we accept that this preventive detention is acceptable in the one case we should do so also in the other.

The threat to oneself The sections of the Mental Health Act which deal with compulsory admission to mental institutions refer both to the 'protection of other persons' and to the protection of 'the patient's health or safety'. Why should mentally disturbed people be *forced* to go into mental hospitals for their own protection? We have already alluded to the countless people who do all sorts of dangerous things which are hazardous to health and safety, and yet they are not forced to go into mental hospitals for their own protection. If these provisions of the Act are not unjust to the mentally disturbed, there must be a relevant difference between them and 'normal' people. But is there such a difference?

Those who attack compulsory detention for the patient's own health or safety often cite the quotation from John Stuart Mill, which appears here on p. 30. However, they usually fail to mention Mill's succeeding remarks:

> It is, perhaps, hardly necessary to say that this doctrine is meant to apply only to human beings in the maturity of their faculties. We are not speaking of children, or of young persons below the age which the law may fix as that of manhood or womanhood. Those who are still in a state to require being taken care of by others, must be protected against their own actions as well as against external injury . . . Liberty, as a principle, has no application to any state of things anterior to the time when mankind have become capable of being improved by free and equal discussion. (Mill 1859, p. 73)

According to the defenders of the law on compulsory detention for the patient's own health or safety, the crucial difference between the patient and other people who act in a way which is likely to endanger their health or safety is that the former are either temporarily or permanently unable to run their own lives. Mill argues that his doctrine should not apply to children because, lacking 'the maturity of their faculties', they are not in a position to decide what is and what isn't good for them. People who are detained under the Mental Health Act for their own protection are deemed to fall into this category. Thus the justification for locking them up, and possibly forcing them to have treatment, is that they are suffering from an incapacity to make rational decisions. Because of this inability they are, in effect, like children, and as citizens we have a duty to protect them from themselves, just as with children.

The people who are detained in this way fall into two classes. First of all there are the 'subnormal', whose intelligence is so attenuated that they are likely to be exploited without knowing it, or to go into the middle of a busy road, unaware of traffic dangers, and so on. They are really the most like young children. Restricting their movements is the least problematic from a moral point of view, because if they were left alone they would almost certainly get killed, or exploited, or perhaps die of starvation, none of which they would want.

The other group is the so-called 'mentally ill'. This class is not defined in the law, but, as suggested earlier, includes people with severely irrational beliefs and those with severely irrational desires. Let us now look at them a little more closely.

What are severely irrational beliefs? Why should having them be enough to justify the compulsory detention of people for their own safety or protection? The first question is harder to answer than the second. The problem of distinguishing rational from irrational beliefs is discussed in the philosophy of science chapter. As you will see when you read that chapter there is widespread disagreement amongst philosophers over standards of rationality of belief. However, there is some consensus that a rational belief is one which is more likely to be true than an irrational one. Some people believe the wildest things, flying in the face of reason. Such a belief might be that one is invincible, or that one is Henry VIII, or worse still, that one is an orange. What is difficult is to define precisely what these beliefs have in common, which makes them all irrational. We shall pass by this problem, partly because it is too complex to be dealt with here, and partly because it is not crucial to this particular debate.

What is special about someone who has radically irrational beliefs of this kind is that he is likely, unwittingly, to get into all kinds of dangers. The person who believes he is indestructible might well walk into the middle of a busy road, as would the subnormal person; he would not be moved by one's reasoning. He would most likely behave in a way which was against his interests, against what he wanted. If someone is prone to having this sort of belief he is unlikely to be able to achieve his goals, whatever they are. The intervention is not an attempt to impose a set of values on him, but simply to help him to be able to act in a more autonomous way, so that he has a better chance of realising his ends.

What about irrational desires? David Hume, the eighteenth-century British philosopher, believed that while one could be said to have irrational beliefs, it didn't make sense to talk of irrational desires – unless by this one meant 'desires which stem from irrational beliefs'. He wrote:

> 'Tis not contrary to reason to prefer the destruction of the whole world to the scratching of my finger. 'Tis not contrary to reason for me to chuse my total ruin, to prevent the least uneasiness of a . . . person wholly unknown to me. 'Tis as little contrary to reason to prefer even my own acknowledg'd lesser good to my greater, and have a much more ardent affection for the former than the latter. (Hume 1738, p. 416)

Hume's point is that if someone has a peculiar want or desire, this does not show that he has made an error of reasoning, and so he is not necessarily suffering from a defective reasoning faculty. Just as some people prefer light-hearted comedies to heavy tragedies, so some prefer one life-style, others another. As long as you know what you are doing, as long as your beliefs aren't irrational, you are not barred by reason from having any set of tastes or desires.

This view would provide a rationale for Mill's views, as quoted from *On Liberty*. As long as someone is able to make decisions based on rational beliefs, he should be allowed to get on with what he wants to do, unless his actions are going to be harmful to others, in which case there might be a justification for preventing him from doing what he wants.

This commitment to freedom is based on the belief that what is distinctive and valuable about human existence is the ability to exercise free will, the capacity to choose one's own fate. To act in a paternalistic way towards someone who has desires which are likely to be self-damaging is to treat him as a sub-human, as a non-autonomous being. This is one reason for not putting people who smoke into mental hospitals, against their wishes, for treatment. With most smokers, at least, it is possible to apprise them of the facts. They might wish to continue smoking despite the risks, or they might wish to give up, but not feel strong enough to be able to do so. But even these latter people would probably prefer to carry on, risks and all, to suffering the inconvenience and indignity of going into hospital for a 'cure', supposing one were available. The racing driver might well prefer the glamour and ecstatic thrills of racing, with all the accompanying dangers, to the more mundane, but safer life which most of us lead. There is a crucial difference between such a person and someone who believes he is Superman and so could never be hurt.

This line of argument would suggest that we should allow the compulsory treatment of people who are suffering from grossly irrational beliefs, but not prevent people from pursuing whatever goals they might have, as long as they don't harm others. However, further reflection reveals grave problems which we shall not go into here. For example, how should we deal with people who have desires which conflict with each other in a self-stultifying way? Suppose that someone is a devoted social worker, and that to be effective he has to appear to the public as a well-behaved, respectable citizen. Most of the time he succeeds. But occasionally he gets an overwhelming desire to get drunk in public. If he does so he will sorely regret it because he wants to be a successful social worker. Would it be right forcibly to prevent him from getting drunk on such occasions? Perhaps the person knows he will desperately regret getting drunk, but at the time is quite prepared to do it. What about someone who is desperately depressed and wants to kill himself? If his depression is based on true beliefs about, say, the state of the economy, but could be relieved by electro-convulsive therapy, enabling the person to live a normal, reasonably contented life, what should you do? If you do nothing he will probably kill himself, and his life will be wasted. If you do intervene compulsorily, at least his life will be saved, but you will have interfered with his liberty.

Another difficulty is in assessing what is meant by simply harming yourself, and what is to count as harming others. A person who smokes himself to death causes great suffering to his family, both because of the grief at his being ill and dying, and in many cases because of ensuing financial difficulties. A person who completely neglects his well-being can be a terrible burden to people around him. Should we regard this simply as a case of someone failing to look after his own interests, or is he to be placed in the category of someone from whom people should be protected? The distinction between behaviour which affects yourself only, and that which affects other people, is crucial, but needs to be thought about a lot before it can be put into practice in cases where decisions about compulsory detention have to be made.

Concluding remarks on compulsory treatment In our discussion of the compulsory treatment of mental illness we have concentrated on whether or not there are significant differences between 'mentally ill' people and others. The point of this is that if it cannot be shown that there are such differences, the special treatment of the 'mentally ill' is arbitrary and unjust. We have argued that there is a variety of different cases, some of which merit special treatment, though others, perhaps the majority, don't.

People who argue about the rights and wrongs of compulsory admission, detention and treatment of mentally disturbed people often take up an extreme position which ignores the strengths of their opponents' arguments. What we have tried to do here, rather than arrive at a neat solution to these very complex questions, is to find some general principles on the subject, which take account of both conservative and liberal attitudes to it. Rather than start out by attempting to prove that one solution is the only or the best one, we have tried to point out the enormous complexities of the moral issues which are involved in drawing up a coherent policy to cope with these problems. Most human problems can't be neatly solved. One must simply make a decision, often to accept the compromise that strikes the best possible balance between conflicting claims.

RIGHTS AND CONSEQUENCES

In the previous sections of this chapter we have talked about the rights and wrongs of various practices which are of vital importance. We have put forward a number of reasons in support of our attitudes to contraception, abortion, infanticide, euthanasia, and the treatment of mentally disturbed people. There is an underlying rationale for the positions taken up, which is common to all the reasons. In discussing these issues we have taken for granted certain basic principles which have lain behind the conclusions we have reached.

All these cases pose serious moral dilemmas because they involve a clash, or at least an apparent clash, between two principles which are fundamental to most of us. Neither of these principles has been argued for here, although people *have,* misguidedly we think, tried to prove that they are true.

The first principle arises from the assumption that an action is right if it benefits people, wrong if it harms them. It is that

> actions are right in proportion as they tend to promote happiness, wrong as they tend to produce the reverse of happiness. (Mill 1861, p. 6)

The doctrine which argues from this base is called 'utilitarianism', and has had much influence on social philosophy for many years.

The principle is important because people's interests often conflict. Frequently, for instance, there is a short supply of goods, and what is to one person's gain is to another's loss. How should such a conflict be resolved? The best solution, according to utilitarianism, is the one which will end up producing the greatest net amount of satisfaction, taking all parties into consideration. Thus, if there was only a limited supply of water available in a country at a particular time, it would be best to distribute it in such a way that everyone had enough to keep him from dying of thirst. One should see to this, even if it meant that some people were denied the pleasure of swimming in their pools. Clearly the loss they would suffer is trivial compared to the benefits to be gained by the people who would die without the water denied to the swimmers.

The other principle appealed to is more difficult to formulate. It is to the effect that it is wrong to treat people in an instrumental way. It is wrong to harm someone simply to benefit someone else. People have a certain dignity, which they have in virtue of being creatures with thoughts and feelings who can make reflective decisions about what they do. Thus, suppose there was a man who lived a miserable life in a town where nobody liked him. In fact his very existence was a source of irritation to most people. Suppose also that he was not able to work productively, and had to be supported by social security all the time. It might be the case, weighing up all the benefits and losses of all those affected, that there would be a net gain if he were killed painlessly in his sleep, compared with allowing him to continue his miserable existence. Many people feel that even if this was so it would still be wrong to kill him. He is a human being, and has a right to be allowed to live. The mere fact that other people would be made better off by his death is insufficient reason for violating his rights.

In this chapter we have not tried to prove that people have rights to life and liberty; nor have we tried to prove that it is right to do that which brings about the greatest net amount of satisfaction. Basic assumptions such as these cannot be proved. If you don't believe that people have rights in virtue of being people, then you will be unimpressed by the arguments, in principle, against abortion, infanticide, and the incarceration of mentally disturbed people. These arguments all rely on the assumption that human beings have rights which should be respected. On the other hand, if you do not believe that it is good to promote satisfaction, you will be unimpressed by nearly all non-religious moral arguments.

We have assumed that these two principles carry some weight with most of the people who read this book, and so we have taken them, in a way, as a starting point. But if someone is genuinely unmoved by these principles we do not claim to be able to prove he is mistaken. So it is fortunate that the interest of social philosophy does not lie in trying to prove basic moral assumptions.

Many people have tried to show that one or the other principle is paramount, and cannot be overridden under any circumstances. The strategy of the utilitarians is to accommodate the rival principle, that one shouldn't violate people's rights, within their framework. For instance, in the case of the man whom people didn't like, a utilitarian would almost certainly try to show that, if you take a very long-term, broad view, including all the side-effects of killing him, you would realise that killing him would not, contrary to what you might at first sight think, bring about the greatest amount of happiness in the circumstances.

The Roman gladiators provide a good example of this. Christians and others were put into an arena in front of thousands of excited spectators, where they had to fight a battle to the death, either with each other, or with a lion. This apparently gave great pleasure to the spectators. Somebody could argue against supporters of our first principle that they would have to agree that the gladiatorial contests were good, because they brought more satisfaction, on balance, than not holding them would have done. There were thousands of spectators and only a few gladiators. So, although the gladiators would no doubt have suffered, their suffering would be outweighed by the pleasure they gave to the spectators. The reply to this is that it is extremely unlikely, all things considered, that more pleasure resulted from the gladiatorial contests than there would have been without them. One should take into account not only the suffering of the contestants, but also that of their families and friends, the insecurity which people, fearing they might end up in the arena, would feel, the diminishing respect which people would have for human life, leading to more acts of violence, and so on. So, after all, the utilitarian would agree that these contests should not have been held.

But the defender of the second principle would reply to this that the utilitarian is dodging the issue. First of all, even if the actual suffering caused by the gladiatorial contests was greater than the pleasure it provided, one doesn't have to go through this sort of calculation to arrive at the conclusion that the contests should not have been held. It is wrong to put people's lives in danger and to cause them suffering just for the amusement of others, even if this would produce a net balance of happiness. Secondly, although the gladiatorial contests probably didn't produce a net balance of happiness, one could construct a hypothetical example where there *was* a net balance of happiness – and this would not make it morally justifiable. Imagine, for example, that the Romans had television all over the empire, so that millions of people could watch the contests. Suppose also that these millions really enjoyed the sport, and there was no substitute available. Furthermore, watching these spectacles did not increase their viciousness. Even if all these assumptions were true, so that there was a balance of pleasure over pain caused by the contests, many people would think they should not be held. This shows that the second principle cannot be incorporated within the first.

Some might balk at the bizarre example introduced above, and also at other examples which philosophers think up. But these examples are not intended to represent actual situations, only to guide one's thought towards the principles which are doing the work in one's judgements about more realistic and complex cases.

If the first principle is not paramount, what about the second? If you believe that the second principle *is* paramount you would be committed to the belief that it is always wrong to kill a person, except, perhaps, when it is his wish that you do so. This fits well with one's intuitions about the gladiators and about the unpopular man. It would also rule out as immoral attempts to create a society like that in *Brave New World*, where the people are like happy automata, with no freedom. But could there be situations where these basic human rights should be overridden? We think there could, although most cases are so complex that it is very difficult to tell when they should be. So, just to show how the second principle ought sometimes to be overridden, we shall present another oversimplified and bizarre example to make the point clearly.

Suppose someone has managed to cultivate a fatal germ against which there is no known cure. He says he will release the germ in New York City unless his brother in Los Angeles, whom he hates, is delivered to him, dead. Suppose also that you can be quite sure that the man will destroy the lethal substance if his brother is brought to him, and he'll definitely carry out his threat if he isn't. What should you do, as, say, a director of police, who could have this innocent man killed? To kill the innocent brother is certainly to violate his right to life, and so goes against the second principle. However, if you don't do this millions will be killed, and the suffering will be far greater. Most people would think it right under these circumstances to kill the innocent brother – as long as you had first tried all methods of negotiation, trickery and stalling on the would-be killer. Not to do this would result in many millions of people dying, whereas complying with the demand would mean that only one person died. If you stick out for the paramountcy of the second principle you would, in this situation, be committed to not killing the innocent brother, and allowing the other brother to kill millions.

We hope that this discussion has shown convincingly that the two principles are independent of each other and are both important. One can neither eliminate one of them, nor reduce one to the other. Attempts to do this fail. But is it possible to produce a rule which will help one to decide, in cases of conflict, which principle should take precedence over the other?

A plausible candidate involves a distinction between harming someone and benefiting someone. We have moral duties not to harm people, which can be called 'negative duties'; and 'positive duties', to benefit each other. When there is a clash between negative and positive duties, the negative ones should take precedence. If there is a conflict between two negative duties the right thing to do is that which does the least harm; if there is a clash between two positive ones you should do that which produces the most benefit. The principle of the priority of negative duties enables one to attach weight to the first principle while ruling out as wrong the gladiatorial contests, and killing the unpopular man. The people who are thrown to the lions are harmed, whilst the spectators are merely benefited. Throwing the gladiators to the lions is wrong because it breaks one's negative duty to the gladiators for the sake of mere positive duties towards the spectators. On the other hand, in the New York City case it could be argued that there is a conflict of negative duties, and that ghastly though it is to kill the innocent man, it is a lesser evil than allowing millions to be killed. What is at stake is the death of one versus the death of millions. At least one must die, possibly millions will die.

The real difficulty with this distinction is that there is no way of distinguishing in the problematic cases between harming someone and failing to benefit him, or between failing to harm someone and benefiting him. In the New York City example it *could* be argued that the right thing to do, given that one accepts the priority of negative duties over positive ones, is to let the millions be killed rather than to kill the innocent brother. If you kill the brother you are harming him, thus failing in your negative duty towards him. If you allow the killer to murder the millions, you have merely failed to benefit them – you haven't harmed anyone.

To focus on the problem consider another case. Is giving food to someone merely fulfilling a positive duty to help people, or is it satisfying a negative duty not to harm people? Taking a friend out for a slap-up meal seems to be a clear case of benefiting someone. But what if the person is starving, and you deliberately deny him food? One way of looking at it is to say that if you deny him food you are harming him, since he will die. But by the same token, if you give him food, he will, presumably, benefit enormously. It seems that the alleged two types of duty collapse into one here. A positive duty is as inseparable from a negative one as is one side of a coin from the other.

But there is *something* in the distinction. In fact there are two important components. The first is the belief that before you concentrate on making some people really well off, it is a higher priority to ensure that other people are not thereby going to be made very badly off. This is part of a principle of equality. For the Roman spectators, watching the gladiators is a luxury, which they should forgo if it means that the gladiators, few of them as there are, have to suffer abject misery. Making their lives tolerable is a higher priority than making the bloodthirsty spectators that extra bit more satisfied.

The second component may be in the minds of people who would argue in the New York case that it would be wrong to kill the innocent man even to save the millions, on the grounds that this would involve violating the negative duty not to kill anyone. What underlies this is a belief that there is an important difference between deliberately doing something with harmful consequences, and deliberately refraining from doing something, with similar harmful consequences. Of course it is true that in the New York City example, if you don't kill the innocent man you haven't actually done anything to anyone, while if you do, you have; but is this morally significant? If the distinction between doing and omitting to do something is itself to provide a basis for deciding what should be done when there is a conflict of interests, it should be applicable to all cases of this type. The following example was suggested by Jonathan Glover, and casts severe doubt on the importance of the distinction.

Suppose a doctor's father is in hospital, suffering from heart disease. The only way he can be kept alive is if a lung ventilator is switched on at a certain time in the morning. The doctor would like his father to die soon, because he will inherit a lot of money from him, and anyway, they don't get on well. At eight o'clock one morning he notices that it is time for the ventilator to be switched on. Let us suppose that he goes into the room, and discovers that the machine has already been switched on. He goes to the machine and switches it off. His father dies. Compare this with a similar situation where he finds that the machine has not been switched on, but deliberately leaves it switched off, and deliberately fails to tell anyone else that it is off. His father dies. If you think there is a significant moral difference between a deliberate act with harmful consequences and a deliberate omission with similar consequences, then you would have to say in this case that in the second variant of our story the man was not as morally culpable as he was in the first, because in the second case he merely failed to act.

Usually when one compares a deliberate act with a deliberate omission that has the same foreseen consequences, there are other factors present which lead one to suppose that there is a moral difference between the two cases. The above example isolates the distinction between acts and omissions, and shows that in a pure form, by itself, it is not morally significant. Perhaps the main feature which is usually present in deliberate acts and usually absent from deliberate omissions is the desire to achieve the consequences. Thus there may be grounds for distinguishing a person who merely does nothing about people starving from someone who actually sends them poison food parcels, not on the grounds that one acts whilst the other merely fails to act, but rather because only the latter actually wants the people to die.

CONCLUDING REMARKS

In this chapter we have tried to show how philosophical arguments can yield conclusions on substantive moral issues, but we do not attempt to prove that our fundamental assumptions about what is right or wrong are correct. The method of discussion is to try to start with common ground, something we hope will be agreed on. Then we look at the broader implications of accepting these assumptions. The process of doing social philosophy involves looking at the relations between the different attitudes one has towards problems in society. The aim is to work out a coherent set of values which reflect most of one's basic intuitions about what is right and wrong. This is why one can use the simplified examples to test theories and principles. If accepting as paramount a moral principle such as that it is always wrong to kill someone can be shown to have consequences that it is wrong to kill even someone with a painful terminal illness who wanted to die, or that it is wrong to kill one person to save ten million, you might give up the principle as such, or modify it to exclude these consequences.

In the last section we discussed two very basic principles. They are that one should try to provide the greatest possible benefits for people, and that one should not treat people as instruments for the satisfaction of others. It seems hard to find a small set of rules which will yield answers to every case. But this should not deter us from investigating different moral principles.

Our discussion suggests that one will not find a simple overarching moral or social principle which can tell one what is right in every case. Principles are set up which more or less work. They are useful as rules of thumb, but should not be regarded as sacred cows. This applies not only to moral principles, but also to the intuitions which underlie them. Although the principles are to be tested against intuitions, each intuition should not be regarded as infallible. They can be tested against one another, and sometimes changed. Social philosophy does not provide final conclusions or ideal theories. But there are better and worse ones. By scrutinising our principles and intuitions, and the reasons for them, we can hope to work towards a more coherent and reasoned set of values about these matters of great concern.

2

Philosophy of mind

'He doesn't know his own mind.' 'Her body's just about packed up, but her mind's as lively as ever.' He thinks he's ill, but it's all in the mind.' 'He did it only with a supreme effort of will.' 'He did it of his own free will.'

These expressions crop up in our everyday lives. Normally they don't cause much difficulty, and we know, more or less, what they mean. They seem straightforward enough, but appearances can be deceptive. As soon as we reflect on what expressions such as these are actually saying, the straightforwardness can rapidly give way to puzzlement. We quickly encounter questions like 'What is a mind?', 'Is the mind something more than just a part of a body?' 'What is an effort of will?'

One particular concern of philosophers of mind is to ascertain whether, and if so, how, the mind can be studied scientifically. As such it is important not only to a person who is puzzled by questions such as 'How do I know that what goes on in my mind is anything like what goes on in someone else's mind?', but also to psychologists. Psychologists have tried to develop a science of the mind. In order to do this a psychologist must have some conception of what a mind is. Otherwise he wouldn't know what he was trying to produce a science of. Psychologists have very different conceptions of the mind. Their contrasting views have led them in quite different directions in their psychological pursuits. One thing philosophy of mind does is to scrutinise the many different frameworks for the scientific study of the mind.

This is a task which, for good reasons, is seldom undertaken within the discipline of psychology itself.

Philosophical problems typically arise when you're in a position where you feel drawn to say one thing about a situation, but are equally drawn to think that this can't be right. Different, but equally plausible, lines of thought lead you to contradictory conclusions, which cannot both be true. Never is this more evident than when we begin to ask questions about the human mind.

In this chapter we shall concentrate on just one of the central problems in the philosophy of mind. It is a problem with far-reaching ramifications for the whole of the philosophy of mind – and beyond it as well. It is the question known as 'the mind-body problem'.

THE MIND-BODY PROBLEM

How are a person's mind and body connected? Does each person have, as well as a body, a non-physical mind, which inhabits the body and directs its behaviour? Or is the mind no more than some part of the body, say the central nervous system? The problem of the connection between the mind and the body is highly controversial as well as having direct bearings on many other questions in the philosophy of mind.

We shall discuss the three main theories about the relation of the mind to the body, each one of which has its supporters today. Although we incline more to one of the theories than to the others, we make no pretence of having settled these age-old problems, and we hope to convey some of the attractions as well as some of the difficulties of each theory. One of the perplexities caused by the mind-body problem is that none of the theories seems to be quite right, and yet they seem to be the only plausible accounts of this puzzling relation.

The three theories are: *dualism*, or the theory that the mind is a non-physical thing, which somehow inhabits the body; *behaviourism*, which is the view that to have a mind is nothing more than to be disposed to behave in certain complex ways; and finally the *mind-body identity theory*, which is the view that a mental state is the same thing as a physical state. Only the first of these, dualism, claims that the mind is something whose existence is independent of bodies. It is the view which seems to reflect common sense most directly. So we shall begin our discussion by considering dualism.

DUALISM

One of the most famous dualists is the seventeenth-century French philosopher, René Descartes. Because of his close association with the doctrine it is often referred to as Cartesian dualism.

In his *Meditations* Descartes sought to establish a firm basis for his beliefs about the world. He wanted to set up sound principles which would enable him to distinguish clearly between genuine knowledge and mere beliefs which might prove to be false. He thought the best way to achieve this was to doubt everything that wasn't absolutely certain, and to see what he could build with what was left. He considered the possibility that most of what he normally took for granted might be false. He noticed in particular that often, when he was actually asleep, it seemed to him in his dreams as if he was awake and doing things which he wasn't doing at all. Also, he noticed that physical objects often turn out to be quite different from how they first appear. He concluded that the knowledge he obtained from seeing and hearing was unreliable. To try to purge himself of all unreliable beliefs Descartes resolved to doubt everything it was possible for him to doubt. As part of this endeavour he supposed, what on the face of it seemed at least possible, that the world was created by an evil demon:

> I shall then suppose, not that God who is supremely good and the fountain of truth, but some evil genius not less powerful than deceitful, has employed his whole energies in deceiving me; I shall consider that the heavens, the earth, colours, figures, sounds and all other external things are nothing but the illusions and dreams of which this genius has availed himself in order to lay traps for my credulity; I shall consider myself as having no hands, no eyes, no flesh, no blood, nor any senses, yet falsely believing myself to possess all these things. (Descartes 1641, vol. 1, p. 148)

Descartes believed that he might be mistaken about all these beliefs, which include the belief that his body exists. It makes sense, according to Descartes, for someone to suppose that his body doesn't exist – that his beliefs about his body are no more than illusions. However, even if there is an evil genius who spends all his infinite power deceiving me, there is one thing I can be quite sure of, namely that *I* exist. I can know this, because even if I doubt everything else, I must exist to do the doubting. It is thus impossible for me to doubt that I exist. But if this is true, what does the 'I' refer to? What am I? What is a self?

Descartes' answer is that 'I' am a thing that thinks. I am not the same thing as my body, or a part of my body, according to him, because it is possible for me to doubt that my body exists without being able to doubt that I exist. I am a thing that thinks and has experiences. I can know that I am thinking and having experiences without knowing anything about my body, and in particular about the state of my brain. So, according to Descartes, I and my thoughts and experiences are something other than a body and its physical goings-on.

On this theory, then, there are many different thoughts and experiences which belong to a non-physical thing called 'the self', which in some rather mysterious way inhabits the body. The Oxford philosopher Gilbert Ryle, who was a staunch opponent of dualism, called this the doctrine of 'the ghost in the machine'.

Self-knowledge and dualism The argument from the fact that I can be sure that I exist without being sure that my body exists, to the conclusion that it is possible that I exist and my body doesn't, and hence to the conclusion that I am something other than my body, is only one route which might attract you towards dualism.

Another route, which also has to do with self-knowledge, is this. Try thinking about something, and then reflect on your thinking. What are you thinking? How do you know what you're thinking? The question 'How do you know what you're thinking?' seems odd because one knows straight off what one is thinking, simply by, as it were, inspecting one's mind. Similarly, if you have a severe pain, ask yourself how you know you have one. Other thoughts and feelings are comparable. Are you sad? Are you angry? Would you like to be somewhere else?

There is something special about knowledge of one's own mental states. I know that I don't have a headache right now. I know that I am sad about the death of a close friend. In order to know these things about myself I don't need to do any research or make inferences from my observations. I seem to know them directly, simply by, as it were, looking inward, by what is sometimes called introspection. There is a peculiar kind of certainty to my judgements about the contents of my own mind.

This certainty does not extend to beliefs about one's body. You can know directly that you have a searing pain, which is apparently coming from a tooth, but you may have to make inquiries to discover the cause of the pain. It might be an abscess, or simply a hole, or it might even be psychosomatic – so that nothing is wrong with your tooth at all! If there's nothing wrong with your tooth, you haven't perhaps, strictly speaking, got toothache, but nonetheless you still have the pain – the pain is real enough, even though it might not have the normal physical cause of such pains.

Thus, it now seems that not only can you know in this special kind of way *that* you are thinking, but you can also know *what* you are thinking and *what* you are feeling. On the other hand, you cannot know directly how your body is. When you know that at this moment you have a pain or that you are now thinking about the Queen, it doesn't seem to involve knowing facts about the state of your body. If you feel sure that the glass of water you see before you is not a hallucination it might not be absurd for someone to argue with you that you are mistaken. But if you are sure that you are in pain or that you feel sad, it would seem to be absurd for someone to argue that you are mistaken. Of course, you might not be able to speak English properly, and so call 'pain' what in English we call 'tickle'. In this case when you say 'I am in pain' you would actually be wrong because you are not in pain, but are experiencing a tickle. But this is a purely linguistic mistake, not a mistake about what you are feeling. You know what you feel, but you use the wrong expression to describe your feeling.

We might know, through scientific research, that headaches are caused by pressure on certain nerves, but it does seem possible that someone could have a feeling which is just like the one caused by this sort of pressure, and yet for there to be no such pressure. To have a thought or a feeling seems to be quite a different sort of thing from your body being in a certain condition.

This peculiarity of our knowledge of our own mental states seems to lead us straight into dualism. Before discussing whether or not it *should* induce us to accept the dualist account of the mind-body relation we shall introduce another line of thought which attracts people to dualism.

Human freedom and dualism One charge often levelled against the Soviet Union by western writers is that people in the Soviet bloc are not free. Whether or not these allegations are true is not our present concern, although it is a very important question. What concerns us is the notion of freedom. When people talk of political freedom they mean something like being able to think and say and do what you want. Freedom in this sense involves not being restricted by other people. But why is this freedom important? This question may sound strange because we all assume freedom is important. But the fact that we don't normally question something doesn't mean that there aren't important questions about it that we should ask.

We seek political freedom only for human beings. We think it quite inappropriate to talk of political freedom for animals. This is because they lack a certain fundamental capacity without which political freedom is impossible. What is this capacity? Human beings are able to deliberate and make up their own minds what to do. We are apparently able to make free choices between alternative actions. Are we unique in this?

Ants join together in cooperative ventures which are of great importance for the survival of ants. If a colony of driver ants has the need to cross a stream, it will set off in a long column. The ants which reach the stream first will march straight into it and drown. As more and more ants drown in the stream their bodies will pile up until finally they form a bridge which the other ants pass over in safety.

Human beings too sometimes lay down their lives for their fellow creatures. We hear from time to time, for instance, of people going back into a burning building, risking their own lives, to save people who may be trapped inside.

We call this human behaviour 'brave' and we praise it, but we don't praise the ants for their 'selflessness'. Rather, the behaviour of the driver ants is, to many of us, frightening, and rather repulsive. Does this merely reflect a bias in favour of our own species, or is there a relevant difference between a brave person and a 'brave' driver ant?

If there is an important difference it seems to be that the brave person acted freely. When a man runs back into a burning building to save someone who might otherwise burn to death he could have done otherwise. If he had wanted to, he could have done nothing, or simply cried for help. On the other hand the driver ants cannot choose what they do. A driver ant is an organism which is, as it were, programmed to behave in certain specific ways. A driver ant at the front of the column couldn't think to itself 'I don't want to lay down my life for these other ants, I'm going to get the hell out of here and save my own skin.' However, when people do things, it seems that they could at least sometimes have done otherwise.

Similar considerations apply to antisocial acts. When a rat attacks another rat, or when a rat attacks a human being, it would just be inappropriate to cast moral blame on it. Rats cannot help what they do, because they are not able to reflect and make up their minds which of the alternatives available to them they are going to adopt. On the other hand a human being who plans and executes a bank robbery probably did think – or at least could have thought – about alternative courses of action, before opting for the robbery. In this case it would be appropriate to condemn him morally.

This freedom which human beings seem to have, and which animals lack, is thought by many to be one of the most important characteristics of being human. A belief that we are free is a precondition for there being any moral or practical problems. It is only on the assumption that we can act freely that questions like 'What should I do?' make sense. If people were not free, at least some of the time, to decide what to do, these questions would simply be inappropriate. But are we really free, or is this freedom just an illusion? According to science all physical objects are bound by the laws of physics. Physical events are connected by causal laws, so, given that one set of phenomena occurs, it is inevitable that it will be followed by some particular set of phenomena rather than another. For instance, if a stone is hurled against a window from close range, the window will break, and if you run an electric current through water, hydrogen and oxygen will be given off. These causal laws also apply to more complex physical mechanisms such as clocks, plants, animals, and the human body. For example, if a human body is deprived of oxygen for a certain length of time, it will suffer brain damage and eventually die; a human body which is infected with a virus will attack the virus with antibodies. Human limbs are all operated by physical mechanisms involving the brain and other parts of the body.

The brain too is a physical object, albeit a very complicated one. Perhaps the reason that no one has ever been able to develop a successful science which could predict the behaviour of human beings is simply that human beings are such complex physical mechanisms that we haven't yet discovered the requisite causal laws. In principle, it seems, we could do this, and one day we might actually do so. Human bodies, although they are human, *are* nonetheless bodies, and all bodies are subject to causal determination. How a body behaves is determined by its condition immediately before it does something, and this condition is determined by its previous conditions, and what other things in the world have done to it, and so on without limit. In a mechanistic universe there is no place for freedom of bodies. So human beings, inasmuch as they are bodies, cannot be free.

And yet people *do* feel that they are free. Imagine that there is a general election with a secret ballot. You have traditionally been a Labour Party voter. However, on this occasion you are dissatisfied with the way they have been handling the economy. After giving it much thought you eventually decide, when in the ballot booth, that the best thing to do is to destroy your ballot paper as a protest. This is what you do. Nonetheless, you *could have* chosen to vote Labour or Conservative right up to the moment you acted. In cases like these, and on numerous other occasions, people make choices which seem to them not to be predetermined.

Here we have a classic philosophical problem – the problem of free will. On the one hand, when we look at human beings as part of nature, from a scientific perspective, it seems that human freedom is impossible, since all our actions are bodily actions, and human bodies are part of an interconnected physical universe in which everything that happens has to happen. On the other hand, when we reflect on our own deliberations and actions it seems clear to us that we do exercise free will, at least sometimes, and that we do sometimes act freely. We can, on some occasions, control what our bodies do.

One apparent way out of this dilemma is to subscribe to dualism. Dualism, we recall, is the theory that a human being consists of two components – a mind and a body. Although they are intimately related, these components *are* separate things which could, theoretically, exist apart from each other. The body without the mind is just like any other piece of matter in that it is subject to physical laws which it cannot alter. On the other hand the mind is able to act, at least sometimes, independently of physical laws. A free action occurs when a non-physical mind acts on a physical body.

Dualism under scrutiny There is more than one test to which we could subject dualism in order to ascertain whether or not we should believe it. Dualism is, as we have argued, supposed to provide an explanation of how human freedom is possible in the face of a physical world which is bound by causal laws. It is also supposed to account for our special kind of knowledge of our own mental states. The test we shall apply here is to see whether or not dualism fares better in explaining these phenomena than its rivals, which are simpler. You will find out in Chapter 3 that it is one of the principles of scientific practice that, confronted with two otherwise equally satisfactory rival explanations for a set of phenomena, we should accept as true the simpler one. If dualism cannot provide a better explanation of the phenomena it was brought in to explain than its simpler rivals, one ought to accept one of them in preference to it.

Let us consider first the phenomenon of one's special relationship to one's own psychological states. Remember the argument that although I can doubt all sorts of things about my body, I can never doubt that I am thinking. I can, apparently, even doubt that my body exists, but I can't doubt that my mind exists. Similarly, if I have a severe pain I cannot doubt that I am in pain, although I can doubt, for instance, that the nerve immediately below my lower right canine tooth is inflamed. If my mind were nothing more than my body, so the argument goes, it would not be possible for me to be certain of all these things about my mind while being uncertain about the state of my body.

Many philosophers would reject the claim that I can doubt that my body exists, and even that I can doubt that there is something wrong with my body when I am in pain. The arguments for both rejections are too complex to discuss here, but they are important for philosophy of mind, and are the focus of many disputes within the subject.

In any case, even if the combination of these doubts and certainties is possible, the reasoning used in the argument is not valid. It wouldn't follow from the fact that I could doubt whether my nerve is inflamed without being able to doubt whether or not I was in pain, that being in pain must be something more than being in a certain physical state. To illustrate this, suppose that two people are looking out of a window at a storm. There is a blinding flash of lightning. One says to the other 'Do you realise that a flash of lightning is nothing more than an electrical discharge?' He replies 'Well, I can't doubt that that was a flash of lightning, but I can doubt whether it was an electrical discharge. Therefore lightning cannot be just an electrical discharge.' Nevertheless lightning is nothing more than an electrical discharge. This shows that the form of argument is invalid, because it can yield a false conclusion from true premises. So one should not accept a conclusion, such as that being in pain is something over and above being in a certain physical condition, just because it follows from such an argument.

But even if you can't prove in this way that the mind is something other than the body, there is still the puzzling fact to explain, that one does seem to have this special kind of knowledge about one's own mind. If you can have a special kind of knowledge about your mind, then perhaps this is because your mind is a special kind of thing. The lightning example is different from that of the mind in that it might be at least *possible* for someone to doubt that, say, there has just been a flash of lightning in the sky, but it does not seem to be even possible for someone to doubt that he is thinking, or if he has a severe pain, that it hurts. And it is at least possible to doubt what sort of physical state one is in.

The answer to this problem is that dualism is not required to explain these facts. The reason it is impossible to doubt that you are thinking is to be found in the meaning of the words 'doubt' and 'think'. It is impossible to doubt that I am thinking because doubting is simply thinking in a certain kind of way. This doesn't show anything about whether or not thinking is more than a physical process. Similarly, the fact, if it is a fact, that I cannot doubt that I have pain when I have pain does not show anything about whether or not pain is just a physical process. Part of the meaning of 'I am in pain' includes 'I am aware of pain', and awareness is a kind of knowledge – the kind of knowledge one has of one's own mental states. Whether these mental states are something over and above physical states is another matter.

What we can conclude from this is that the apparent pressure towards accepting the relatively less simple theory of dualism, which comes from our direct awareness of our own mental states, can be resisted, once we realise the reason why this awareness is direct.

The second major pressure towards dualism comes from the belief that we are free, and that this is possible only if we have non-physical minds which can act independently of physical causes.

In order for dualism to help solve the free will problem we would have to be able to explain just what it means for an action to be free, and how such an action differs from a mere physical happening in the brain. Let's take as an example the difference between someone having a real epileptic fit and someone in a play merely pretending to have one. If the second person is a good actor there will be very little difference between his observable bodily movements and those of someone who is having a real fit. For instance, similar facial muscles will be contorted in both cases.

Let us focus on one contortion of the facial muscles around the mouth. In the first case these muscles are contorted in an uncontroversially deterministic way, and the person having the fit could not do anything about it. He did not choose to contort his muscles. It just happened, and was inevitable given the state of his body immediately before it happened. The second person, however, chose to contort his muscles, 'of his own free will'. It was not, apparently, inevitable, given the state of his body just before he started twitching, that he would do so. He made a decision that this was what he was going to do, and then he did it. But how?

He made the expressions on his face by moving certain muscles. The pull on the muscles caused the skin around the mouth to tighten. But what pulled the muscles? An impulse from the brain. But what caused the impulse from the brain to be emitted? The answer a dualist would like to give is 'the mind' or 'the will'. The non-physical person non-physically willed his physical facial muscles to contort. The person willed his muscles to move and they, with the help of the physical brain and the physical nervous system, as it were, obeyed his command.

But this kind of answer raises enormous difficulties. First, how is the will, which is by definition a non-physical thing, able to have effects on physical things? Physical things have effects on each other. But what could it mean to say that a mind caused a body to behave in a certain way? It's all very well to talk in a metaphorical way about the mind acting on the body, but this isn't much good unless the metaphor can be explained.

A second difficulty is that the dualist thesis that on some occasions the non-physical mind causes the body to behave in a particular way through mental intervention is actually incompatible with the view that *all* physical bodies are causally determined by physical laws. If physical determinism is true, then how any body behaves is determined by its immediately preceding state and the physical actions on it of other determined bodies. Thus the apparent reconciliation of freedom with determinism is no reconciliation, but just a denial of determinism.

The dualist might of course say that determinism is true of all bodies except, under some circumstances, human ones. If this is so it would be a puzzling fact that human bodies were unique in their lack of physical determination. The dualist can apparently explain this by saying that human bodies are different simply because they are inhabited by non-physical minds. But it would then be incumbent on the dualist, if his explanation is to be a genuine one, to be able to say more about the non-physical properties of minds, which enable them to act in this way. Unless the dualist can do this his 'mind' is nothing more than 'that thing, whatever it is, which makes human bodies different'. As far as this argument for dualism goes, what is peculiar about human bodies might just as well be something physically special in their make-up.

There is a further difficulty for the dualist's supposed solution to the free will problem. According to the view under discussion, when someone moves a part of his body the whole process is started off by an action of the mind. The person wills that something happens, and this causes it to happen. But we still might ask: What caused the initial act of the will? It seems that there are two possibilities here, neither of which helps the believer in human freedom. Either the action of the mind was caused by something else, say another mental action, or it just happened spontaneously, without any cause.

Suppose, for example, that someone decides to bang his fist down hard on a table. What triggers off the physiological reaction is, according to dualism, an act of will. But if this act of will just happens, then the person is not in control of it, and he can hardly be said to be free. The other alternative is that it doesn't just happen, but happens for a reason, say that the person wants to emphasise a point he is making, in which case we can ask questions about why the person wants to emphasise the point. The answer to this might be that he wants to emphasise the point because he wants it to be noticed, and believes that if he doesn't emphasise it it will pass unnoticed. Eventually we can always trace what people 'freely' do to certain desires or goals, and certain beliefs.

If the dualist is to make his case he would have to show that the desires and beliefs which motivate people are freely chosen. Again there seems to be a dilemma. Either desires and beliefs just spontaneously happen, or they come about as a result of something else. If the former were true they would be random, if the latter they would be determined. Neither seems to be compatible with freedom. It is just as difficult to explain how a dualist's 'act of will' could be free, as it would be for a behaviourist or an identity theorist to explain how human freedom is possible.

The idea of regarding the mind as a separate entity from the body was introduced as a way of explaining how people can be free, given that the physical world is a deterministic one in which everything that happens is caused by something else. But saying that, when people do things, what they do begins with some mental, non-physical, event, doesn't solve the problem at all.

The problem of accounting for our widespread belief that we are free agents, in the light of our equally widespread belief that the physical world is a deterministic one, is one of the most intractable problems in philosophy. We shall not be foolhardy enough to attempt to solve it in these few pages. What we do hope to have shown is that, contrary to first appearances, dualism, the more complex theory, can no more solve this puzzle than can its rivals.

There are many reasons why dualism attracts people as a theory of the relationship between mind and body. We could not possibly discuss here all the reasons for believing it true. What we have argued is that some of the leading reasons for believing it do not, under scrutiny, stand up as good reasons.

BEHAVIOURISM

Consider the following bizarre passage from a story by Lewis Carroll:

> 'But oh agony! Here is the outer gate, and we must part!' He sobbed as he shook hands with them, and the next moment was briskly walking away.
>
> 'He might have waited to see us off!' said the old man piteously. 'And he needn't have begun whistling the very *moment* he left us!' said the young one severely. (Lewis Carroll, *A Tangled Tale*, Knot VIII)

Why is this passage odd? Surely because it is difficult to accept the idea that a person could truthfully say that he felt agony (intense grief) at one moment, and then at the next start whistling. And this is because whistling is not the kind of behaviour which we associate with people who are suffering real grief. The incident in Carroll's story is peculiar because we think that actions speak louder than words.

Could a person who was actually experiencing intense grief whistle the tune 'Oh what a beautiful morning!' in a brisk fashion? It might be said that there is nothing really absurd about such a case. Although normally people who are grief-stricken do not behave like this, we can describe a situation which would make the case intelligible.

Imagine that the man in Carroll's story is real, and that the passage describes an incident which actually occurred. On learning of it we approach the man concerned and accuse him of insincerity. We tell him that it is quite obvious from his behaviour on this occasion that parting from the others was not painful. He protests that on the contrary the grief he felt on parting from the others was very deep and real; and this is why he sobbed. But, he adds, he is by nature stoical, and so he quickly pulled himself together and made an external show of unconcern by walking off briskly and whistling. Later that day we encounter the same man again. He is unaware that we are observing him. He has a dead dog cradled in his arms. He is sobbing aloud that the dog is his, and that it was run over by a car while trying to cross the road. He tells some bystanders that he is absolutely distraught; the dog has been his faithful companion for many years. He sobs for quite some time. What are we to think of this man's claim to have felt real grief on parting from the others now?

We might think that this man is after all a liar; he did not in fact feel anything when he left the others at the gate. The new evidence we have contradicts his claim that when he feels grief he shows no outward concern. Or we might think (rather oddly perhaps) that when he is grieving about *people* he does not display his emotion; but when his grief is for animals he gives public vent to his feelings. If we did say this we should be left wondering why he behaved in this strange way. How can we find out which answer is correct?

We know from our previous discussion what kind of an answer the dualist would give. He would argue that since people are composed of minds and bodies the only way we can really find out if another person feels grief, which is an event in his mind, is to ask him, and for him to answer truthfully. The workings of his mind are not outwardly visible; all that we can observe are the movements of his body. Each of us, however, in virtue of having a mind, has infallible knowledge about the workings of his own mind. I know for sure at any given moment whether or not I am grieving about something or other. And in general, each of us knows what his own thoughts, emotions and sensations are.

Despite the objections to dualism that we have already discussed there is something compelling about this picture; and yet – and this is the odd thing – we don't normally act as if we think it true in our daily lives!

Think again about the man parting from the other men. How plausible did you find his protestation that he really was in 'agony' at the parting? And even if you believed that there was a remote chance that he was telling the truth, did you continue to think this after his behaviour in the case of the dog?

The dualist insists that since each of us knows from introspection whether or not he is in a certain mental state, the incident with the dog shows at most that the man lied to us when he said that he felt grief on parting from the others. The dualist would either have to say that, or opt for the theory that this is a man whose grief has outward signs only when it is felt on account of animals. Now suppose that we gather further evidence about the character of this man. Suppose we discover that he has lost a close relative and that at his funeral service he had exhibited signs of grief as evident as he did in the case of the dog. What would the dualist say then? Again, he could maintain that the man originally lied to us, or he could construct a new theory to account for the new evidence. Perhaps he conjectures that the man only manifests public grief for animals and relatives.

What becomes clear is that no matter how much evidence we gather about this man's behaviour, the dualist will always insist that either the man is not telling the truth about his feelings, or that he is telling the truth but that certain things are true of this man that are not generally true of the rest of us. In general the dualist will always insist that when a person maintains that he is grieving, and yet his behaviour is not the kind of behaviour typical of grieving persons, either he is not telling the truth, or some special theory has to be devised to explain why his claim about what he feels doesn't fit his behaviour. And since (as in the case we have imagined) we may gather further evidence about his behaviour which contradicts the theory that was brought forward to explain the original lack of fit, we could envisage a whole succession of theories of an increasingly complicated kind; and this is a way of saying that for the dualist *no* particular forms of behaviour *need* be associated with grief or its absence. And the same point also holds for other mental phenomena such as feelings of pain.

The behaviourist finds this view quite implausible. It is, he thinks, quite obvious that the man in Carroll's story was lying. Either that, or he does not understand the meaning of the words he uses. The behaviourist maintains that how a person behaves is *the* ground for deciding what mental properties he has. And this is so, according to the behaviourist, because the concepts 'mind' and 'mental property' are in fact concerned only with the ways in which bodies behave; they make no reference to anything else.

Recall that the dualist pictures the mind as a kind of ghostly theatre with an audience of one. How is each of us supposed to find out what play is being performed in any theatre other than our own? This is a deep difficulty confronting dualism and it is referred to as 'the problem of other minds'.

The dualist cannot say that we can know a person is feeling grief because he generally behaves in ways similar to the ways in which we behave when we are feeling grief. He cannot say this because he has given us no way of telling whether the feeling of grief *is* associated with certain forms of behaviour in any case other than in our own. Similarly, if we inadvertently rest our hand upon a very hot surface we may cry out with pain. But what is the justification for our belief that other people feel pain when they place their hands on a hot surface? The dualist is not entitled to say that their crying-out behaviour is a reliable indication that they do feel pain, since he can provide no way of establishing this. We have merely scratched the surface of the problem of other minds here, but it confronts dualism with what seems to be an intractable difficulty.

In the course of our daily lives we confidently ascribe a vast range of mental properties to people. Dualism seems to entail that our confidence is unfounded. The behaviourist resists this sceptical conclusion by giving us all a ticket to each other's theatre; minds become open to public observation.

Two kinds of behaviourism There is an important distinction to be drawn between two kinds of behaviourism. The first kind is often known as 'methodological behaviourism', and it is advocated by certain psychologists who want to study human behaviour from the standpoint of science. The second kind of behaviourism, which is called 'analytic behaviourism', is a philosophical theory advanced by some philosophers hostile to dualism. Our chief concern here will be with the latter doctrine; but first we want to make some remarks about methodological behaviourism, in order to show how the two kinds are related.

One of the best-known proponents of methodological behaviourism is the American psychologist B. F. Skinner. Skinner argues that the only way to construct an effective science of human behaviour is to focus our attention on what we can actually observe. We don't observe people's beliefs, intentions, desires and so on; therefore we should not make reference to them in the general principles which we put forward to explain human behaviour. The following passage from Skinner's book *Beyond Freedom and Dignity* gives us something of the flavour of one of his leading ideas:

> The mental explanation brings curiosity to an end. We see the effect in casual discourse. If we ask someone, 'Why did you go to the theatre?' and he says, 'Because I felt like going,' we are apt to take his reply as a kind of explanation. It would be much more to the point to know what has happened when he has gone to the theatre in the past, what he heard or read about the play he went to see and what other things in his past or present environments might have induced him to go (as opposed to doing something else), but we accept 'I felt like going' as a sort of summary of all this and are not likely to ask for details. (Skinner 1971, p. 18)

Actually, Skinner is not being quite fair here, since if we asked a person why he was going to the theatre and he replied 'I feel like going', far from thinking that he had explained why he was going, we would most likely think we were being rudely told that he had no intention of giving us any explanation! When we ask someone why he did such-and-such we normally suppose that he wanted to do what he did; what we are asking him to supply is an explanation of why he had the want or desire that he in fact had. 'I went to the theatre to see the play by so-and-so, and I regard his work as the most significant in the contemporary English theatre. I think his work is significant for such-and-such reasons.' This would be one form that an explanation of why a person went to the theatre might take, and it plainly goes further than the unhelpful 'I felt like going.'

Skinner's general point, however, is that even the expanded 'explanation' of why a certain person went to the theatre is not really an explanation at all. It uses the words 'think', and 'reason', all of which are specifically mental terms. Only a creature with a mind can be said to think or to regard or to have reasons. Skinner claims that although we naturally try to explain the behaviour of others by appealing to mental goings-on, such talk actually blocks the road to inquiry; in order to build a successful science of behaviour we should

> follow the path taken by physics and biology by turning directly to the relation between behaviour and the environment and neglecting supposed mediating states of mind. Physics did not advance by looking more closely at the jubilance of a falling body, or biology by looking at the nature of vital spirits, and we do not need to discover what personalities, states of mind, feelings, traits of character, plans, purposes, intentions or the other prerequisites of autonomous man really are in order to get on with a scientific analysis of behaviour. (ibid, p. 20)

This then, is the thesis of methodological behaviourism. There exist causal links between behaviour and the environment which are amenable to systematic and scientific study. It is confusing and unhelpful to appeal to states and events of a mental kind going on inside people. This is so because, firstly, these mental events and states themselves have external, observable causes, and so we might as well look directly at them, and secondly, these mental states and events are unobservable except by those that have them, and unquantifiable, which makes them unsuitable ingredients to incorporate into a science of behaviour.

We should notice that methodological behaviourism is not committed to maintaining that people do not have minds and that mental processes are unreal. All that is insisted upon is that reference to such processes will not be made in a science of behaviour. This at any rate is Skinner's official position.

Now let us turn to the philosophical doctrine of analytic behaviourism. This doctrine, at its most general, claims that our talk about minds and states of mind can be translated or 'analysed out' into talk about bodies and their observable properties. In *Beyond Freedom and Dignity* we can find traces of this doctrine as well as of methodological behaviourism, Skinner's official position. He has this to say:

> No one looks askance at the astronomer when he says that the sun rises or that the stars come out at night, for it would be ridiculous to insist that he should always say that the sun appears over the horizon as the earth turns or that the stars become visible as the atmosphere ceases to refract sunlight. All we ask is that he can give a more precise translation if one is needed . . . No doubt many of the mentalistic expressions embedded in the English language cannot be as rigorously translated as 'sunrise' but acceptable translations are not out of reach. (ibid, p. 28, cf. p. 16)

So, just as we can translate talk about the sun rising into talk about the earth turning, so we can translate talk about the workings of people's minds into talk about their behaviour. And just as our scientifically purified translations of talk about the sun will make explicit the fact that the sun does not really rise, so our translations of mind-talk will reveal the falsity of dualism.

If analytic behaviourism should turn out to be true then this might provide a plausible philosophical justification for why methodological behaviourism made good sense in psychology. On the other hand if, as we shall argue below, analytic behaviourism is inadequate as a general account of the mental, then this fact ought to illuminate the contemporary debate between Skinner and his critics about whether the restriction of psychological science to what can be observed may make it impossible to come up with any interesting explanations of human action, since it leaves out of account part of what is to be explained.

The dualist, we recall, argued that the attribution to creatures of mental properties such as anger, intelligence, dreams, perceptions, and so on, presupposes that the creatures in question are creatures with minds, since the mental properties so attributed must be properties of something. The dualist thinks that this something cannot be the body of the creature, since we can see what is happening to people's bodies but we cannot see their mental life; we cannot observe their dreams, perceptions, pains and so on.

The analytic behaviourist argues that the dualist has seriously distorted and misunderstood our talk about the mental lives of one another. He agrees with the dualist that we do attribute mental as well as physical properties to people. But he denies that these facts support the conclusions drawn by the dualist. He denies that it follows from the fact that we attribute both mental and physical properties to people that the subject of mental properties is an immaterial mind, while the subject of physical properties is a material body.

To help in evaluating the claims of analytic behaviourism let us start by imagining a lump of sugar lying on a table. The sugar is dry; it is a cube in shape, white in colour, and has a crystalline structure. We can discover that the lump of sugar has these properties by direct inspection. There are further properties which can easily be discovered by more or less direct inspection if we have reasonable intuitions; for instance we can judge approximate weight and volume. (You may feel impatient here and start wondering what all this has to do with behaviourism; but in fact, as we shall see shortly, consideration of this case will lead us directly to the cardinal claim made by the theory.)

Now sugar also has the property of being soluble, which is to say that it dissolves in a liquid. But is solubility a property of our sugar lump which we see in the same way as we see that it is white or cubical? It seems not. To say that the sugar lump is soluble is, rather, to say that it has a particular disposition; the disposition to dissolve when placed in water.

Let us draw an admittedly rough and ready distinction between observable and dispositional properties of objects. Observable properties of objects can be directly recognised by one or more of the five senses, while dispositional properties cannot. Thus, we can observe that something burns or bends but not that something is inflammable or flexible. So 'burns' and 'bends' refer to observable properties of things; while 'inflammable' and 'flexible' describe dispositions. But what does it mean to say that something has an unobservable property?

Let us imagine a possible but rather odd reply to the question: the fact that direct observations of the sugar lump reveal that it is white, cubical, and so on, but do not reveal that it is soluble, shows that it is actually composed of two kinds of thing or substance. It consists of a physical substance, which has the directly observable properties of the sugar cube, and an immaterial substance, which has the dispositional properties of the cube, which we don't directly observe. Of course, we observe the manifestations of the dispositions of the cube, but there is plainly a difference between observing the manifestations of a disposition and observing the disposition itself. We can't deny this without saying that a sugar lump which never as a matter of facts gets dissolved is insoluble, and this is strongly counter-intuitive. Thus the fact that sugar cubes have both dispositional and non-dispositional properties means that they must be composed of two kinds of thing or substance.

This absurd chain of reasoning gives us the clue to the chief strategy of the analytic behaviourist. Of course, he says, it is absurd to infer from facts about dispositional and non-dispositional properties that sugar cubes consist of two sorts of thing. But this is precisely what the dualist does when he reflects upon our talk of the physical and mental properties of persons. When he reflects on this he makes a mistake no less absurd than the mistake made in the case of the sugar cube. For he supposes that if people think, perceive, want, and so forth, they must be composed of two sorts of things, minds and bodies.

The dualist is led to make this mistake, at least in part, because he thinks that although we do observe a person's bodily properties such as his height and build we don't observe his mental properties such as his intelligence or anger. What we actually see are the manifestations of his intelligence or anger. Also, a person does not lose all his mental properties when he is asleep. For example, a generous person does not cease to be a generous person when he falls asleep.

The behaviourist claims first that mental properties are dispositional properties, and secondly that dispositional properties can be analysed in such a way as to defuse dualism. The correct analysis of mental properties will show that people are not anything over and above bodies.

How then should we analyse dispositional terms? The analytic behaviourist maintains that the most plausible analysis is one which asserts that (for example)

(*a*) This lump of sugar is soluble.

means the same as

(*b*) If this lump of sugar had been placed in water then it would have dissolved.

or

(*c*) If this lump of sugar were placed in water then it would dissolve.

Sentence (*b*) is the translation we would offer for the case of our sugar lump which is never placed in water throughout its life.

The analytic behaviourist believes that this sort of analysis removes the temptation to think that there are ghostly goings-on in sugar cubes. And he also argues that we can offer parallel analyses of mental properties which will similarly remove the temptation to think dualistically about people.

Consider some things which might be said about a certain individual Smith:

(*a*) Smith is intelligent.
(*b*) Smith is surly.
(*c*) Smith hates his mother.
(*d*) Smith wants to make a boat.
(*e*) Smith believes that London is in England.
(*f*) Smith dreamed that he was the King of France.

The fact that we might say any of these things about Smith does not show, according to the theory, that we would be making assertions about a non-physical ghostly 'mind' of Smith; this would be a mistake analogous to the error made in the case of the sugar cube. People are not composed of both minds and bodies, but are simply bodies which behave in very complicated ways. And we are pointing to these complicated ways of behaving when we ascribe mental properties to people. To say that a creature has a mind is to say that it is appropriate to ascribe mental properties to that creature. And creatures with minds are persons. But to ascribe mental properties to a person is not to ascribe an immaterial mind to him, but to say that a material body behaves, or is disposed to behave, in certain sorts of ways.

If we should protest that people are not simply material objects, as is shown by the fact that we don't attribute mental properties to sticks and stones, the behaviourist replies that our objection begs the question. Of course, we do not ascribe mental properties such as intelligence or anger to sticks and stones, and we can, if we like, express this by saying that sticks and stones don't have minds. But this only means that sticks and stones do not behave in the complex range of ways that humans behave; in fact they don't *behave* at all.

The behaviourist also points to the fact that as the behaviour of a creature becomes more complicated, so we feel more inclined to attribute mental properties to it. We are more inclined to say, for instance, that cats have beliefs than we are to say that worms do.

Let us consider how the behaviourist treats the mental properties mentioned on the list (*a*)–(*f*). Remember that the dualist maintains that Smith, like every other person, has introspectively infallible knowledge of his own mind. So he knows whether all the mental properties on the list really do belong to him. In particular he knows whether or not he is intelligent. Now the dualist thinks that while we can only conjecture whether the property of intelligence is true of Smith, Smith himself knows whether it is. Writing more than three hundred years ago, the English philosopher Thomas Hobbes gave cynical expression to this view when he wrote:

For such is the nature of man that howsoever they may acknowledge many others to be more witty, or more eloquent or more learned; yet

> they will hardly believe there be so many so wise as themselves: for they see their own wit at hand, and others at a distance. (Hobbes 1651, p. 184)

But the fact that Smith believes himself to be intelligent does not prove that he is intelligent; for, argues the behaviourist, to say that Smith is intelligent is to ascribe not an occult property to an invisible mind but a set of dispositions to a body.

The set of dispositions which give the meaning of the sentence 'Smith is intelligent' may include such dispositions as:

> If Smith is given the *Times* crossword puzzle to do he usually completes it within an hour.
>
> If Smith is given mathematical problems to do then he solves them quickly and correctly.
>
> If the principles of cost-benefit analysis had been explained to Smith, then he would soon have understood them.

It is impossible to specify all of the dispositions which have to be true of Smith in order for him to count as intelligent. We can't specify them all because the manifestations and potential manifestations of Smith's intelligence depend upon his projects (wants and desires) and on his environment. For this reason, the property of intelligence is said to be a 'multi-track' dispositional property. Although problem-solving behaviour is a part of what it is to be intelligent, we cannot specify a single set of ways of problem-solving (even for the one individual Smith) which constitute intelligence. The behaviourist thinks that all of the mental properties on our list are multi-track dispositions. They contrast with 'single-track' dispositions such as being soluble.

What follows from this view is that a person might well be mistaken in supposing that he was intelligent or that he had a certain belief. Unlike the dualist, the behaviourist claims that other people are often as well or even better placed to determine these things. Nor can we object to this doctrine by pointing out that a person can be intelligent or have a certain belief without giving the slightest indication of these facts to those around him; for he still has the relevant dispositions to behave in certain ways, just as sugar remains soluble even when not placed in water.

The behaviour of people is not just *evidence* for the existence of mind; that very behaviour *is* the working of mind; and, says the analytic behaviourist, if we think about it we will see that in practice we do assess the mentality of others by observing their behaviour. If, for example, a person never performs a single intelligent act in circumstances where it would be appropriate for him to do so, it is useless for him to protest that he is really intelligent. Dualism is an illusion engendered by misunderstanding the facts about mental discourse.

Analytic behaviourism under scrutiny How plausible is the doctrine as a philosophy of mind? It seems to be very plausible for some features of mentality; but unless it can be shown to be plausible for all features of mental life it will be inadequate as a comprehensive theory, although it may be true for certain properties of mind.

The theory seems very plausible with respect to traits of character such as generosity, humourlessness, aggressiveness and untrustworthiness. The leading feature of character traits is that they change very slowly, if at all, through time. For instance, an ungenerous person does not suddenly become generous. (Scrooge in *A Christmas Carol* did just this; but there were admittedly special circumstances attached to his transformation!)

Where character traits are concerned, claims by a person to the effect that he does or does not possess a particular trait are often very suspect. For example, if a person continually breaks his promises in circumstances where there are no special reasons which would excuse his conduct, we are not impressed with his claim that he is 'really' trustworthy. We decide a person's character traits after watching his behaviour in a wide variety of social settings; and the behaviour *is* the criterion for settling whether a person is generous, affable, surly, or intelligent. (That this is so is shown in part by the reflection that people are often prepared to correct their own estimates of their characters on the basis of the judgements of others. But it plainly requires a greater degree of honesty to admit that one is surly than it does to admit that one is intelligent or generous!)

Let us next consider mental states such as wants and beliefs. Can we give an adequate dispositional account of these? Notice that these mental states, unlike traits of character, may change rapidly through time. For instance, a man who awakes with the false belief that it is raining has only to glance through his window at the cloudless dawn to have his belief immediately cancelled. Again, a person who wants the Liberal Party to win the next general election may abruptly discard his desire if he alters certain of his beliefs about the Liberal political programme. Since belief is such a central feature of mentality let us attempt to analyse it dispositionally.

Suppose the following is true of an individual called Smith:

1 Smith believes that London is in England.

We wish to give an analysis of (1) which avoids using the mental expression 'believes that London is in England' and makes reference instead to Smith's actual or potential behaviour. Suppose we try:

2 If Smith were asked the question 'Is London in England?' then Smith would answer affirmatively.

Now (2) is certainly dispositional in form and it apparently eliminates reference to a state of belief – but does it? If I say 'Is London in England?' I am speaking English. Let us suppose that Smith does not speak English. Then, when I ask him this question, he will not understand it, and consequently he will not respond affirmatively! Now you may find it implausible that anyone with a name like 'Smith' could possibly fail to understand English, but that is not the point. The point is that (2) plainly does require that Smith is able to understand English, so we ought to make it explicit by amending (2) to:

3 If Smith were asked the question 'Is London in England?' and Smith knew the English language then Smith would answer affirmatively.

But what is it to know English? A person who knows the English language must know what sentences of English mean. This much seems obvious, but what is far from obvious is that we will be able to give a dispositional account of what it is for a person to know the meanings of English sentences. In general, the concept of meaning is one which is extremely puzzling. Some of the problems connected with the idea of meaning are taken up in Chapter 4. For the present we shall side-step the difficulties involved in the idea of knowledge of a language and pretend it is quite straightforward. Our difficulties, however, would not even then be at an end.

Let us suppose that Smith understands English all right, but he wants to conceal his belief that London is in England from us. Now Smith will not respond affirmatively when we ask him if London is in England. He may even respond by denying it in order to heighten the deception. So (3) needs further amendment to take care of this possibility. Perhaps we should amend it to read:

4 If Smith were asked the question 'Is London in England?' and Smith knew the English language, then Smith would answer affirmatively provided he did not want to deceive his audience.

Is (4) a satisfactory dispositional analysis of (1)? One thing is clear: (4) makes reference to the psychological phenomena of deceiving and wanting. These too need to be given a behavioural interpretation in order for (4) to be accounted satisfactory. Otherwise we should only have succeeded in giving a dispositional account of belief by introducing other psychological states; yet the aim of analytic behaviourism is to explain *all* psychological phenomena in dispositional terms.

Can we give a dispositional account of attempted deception, as we certainly need to do in order to count (4) as a satisfactory analysis of (1)? What is it to try to deceive someone? It is to say something (or to perform some other kind of action) with the intention of causing your audience to acquire a certain belief, or set of beliefs, which are in fact contrary to what you believe. So in order to give a behavioural analysis of attempted or successful deception it is necessary to give a behavioural analysis of belief. But this latter task is just the one we have been engaged upon. Now since our analysis (4) of (1) makes reference to the idea of deception, and since deception turns out to require the idea of belief, it follows that (4) is tacitly circular – and hence objectionable – because it makes covert reference to belief. And belief is what we have been trying to eliminate from (1) in order to fulfil the programme of analytic behaviourism.

In order to carry through the programme of analytic behaviourism we have to translate assertions about the mental states of individuals into statements which make reference to the individuals' potential and actual patterns of behaviour. Our argument suggests that such translations end up by making covert reference either to the mental state which is supposed to be being analysed in terms of behaviour, or to other mental phenomena. We don't therefore succeed in eliminating 'inner' goings-on in favour of observable behaviour. Of course, we have not actually proved that analytic behaviourism is impossible, and it is an interesting exercise to try to remove some of the difficulties which we have dicussed; but all attempts to provide purely behavioural translations of mental phenomena which the authors have seen are circular in the way just described.

So far we have considered two categories of mental phenomena, character traits and mental states. We have concluded that we are able to give a satisfactory dispositional account of the former class but not of the latter. To put the point crudely: there is more to mind than observable behaviour. Some philosophers have been driven to the view that mental states such as beliefs and desires can be given a dispositional analysis because, to take the case of belief, it is absurd to suppose that all of a person's beliefs are at the forefront of his consciousness. We all have plenty of beliefs of which we are unaware at any given time. Also, when I am asleep I do not cease to believe, if I do believe, that the square root of 16 is 4, yet I am certainly not aware of my belief. A person who desires praise may be partly or wholly unaware of his desire for a time. These and countless other examples show, then, that we cannot maintain the view that a person is always aware of his mental states; and, as in the case of character traits, we do often accept testimony from others about our beliefs and desires. But, as we have seen, these facts don't warrant our concluding that mental states are dispositions.

There is another category of mind which we have not looked at, and for which analytic behaviourism seems even less plausible, namely the category of mental events. Mental events are occurrent goings-on which I am immediately aware of, such as the experience of feeling a pain in my foot or the experience of looking at a red object. My whole waking life is packed with these events – and much of my sleeping life too.

It has seemed to many people who think about mental events that it is just self-evidently false that our images and current sensations could be identical with any kind of behaviour. How could the 'felt' difference between the experience of seeing something red and the experience of seeing something blue be explained as a difference in actual or potential behaviour? We do seem to be directly aware of events in our own minds in a way in which others are not, and therefore, it seems, our images and sensations, and all the mental events which constitute our 'stream of consciousness', cannot be read off directly from our behaviour.

The reader may feel that these considerations are powerful and compelling; and that our earlier discussion of analytic behaviourism is therefore unnecessary. It is a fact, however, that many analytic behaviourists remain unimpressed by the appeal to immediate experience. They argue, for example, that while we can say quite a lot about the objects and events of which we are conscious it is remarkably difficult to say anything interesting about consciousness itself. Consciousness itself, by contrast with the things of which we are conscious, seems peculiarly featureless. The analytic behaviourist complains that it is unfair to expect a behavioural account of something about which nothing can be said.

The reader probably inclines to the view that this move is too easy, and we agree. We need an account of mental events. And we need an account of these which does not drive us back into the arms of dualism, which is, as we saw earlier, a very implausible theory. We concentrated on mental states in our discussion of behaviourism because we hoped to show that the dispositional account is unsatisfactory as an analysis of beliefs and desires; and these *are* features of mind about which a great deal can be said.

THE MIND-BODY IDENTITY THEORY

At the beginning of this chapter we said that there are only three plausible accounts of the mind-body relation. Having rejected the first two we come now to the final one – the ***mind-body identity theory***. The identity theory, as we shall call it, shares with behaviourism the view that human beings do not possess, as well as bodies, non-physical minds. Where the two theories differ sharply is in just those areas where we found behaviourism deficient – the analysis of mental states, such as feeling sad, and of mental events, such as being struck by a sudden thought that you are half-an-hour late for an appointment. The identity theory rejects the view that mental states and events are dispositions to behave in certain ways, and claims instead that they are *identical with* bodily states and events.

If dualism, behaviourism and the identity theory are the only three plausible theories of the mind-body relation, and if dualism and behaviourism are false, then the identity theory must be true – and yet on the face of it, it can appear as the least attractive. We think that a number of its apparent drawbacks are due to a misunderstanding about what the identity theory really says. So we shall spend the last few pages of this chapter trying to dispel some common illusions about the identity theory.

The main misconceptions about the identity theory arise from not fully grasping what is meant by saying that mental states are identical with physical ones. What does this mean? This question seems much more straightforward than it actually is. What we mean by 'identical' is best illustrated by looking at another, less problematic, 'identity statement'. For many years astronomers observed a star which occupied a certain spot in the heavens in the evening, and because it had this feature it was called by them 'the Evening Star'. During the same period they also observed a star which occupied a certain spot in the heavens in the early morning, and this was called 'the Morning Star'. After a considerable time, with the help of much observation and calculation, it was discovered that the Morning Star was identical with the Evening Star. To say that the Morning Star is identical with the Evening Star is to say that the Morning Star and the Evening Star have all the same properties. For instance, if the Morning Star is about to explode, then so is the Evening Star. This remark might seem puzzling, simply because it is so obvious as hardly to need saying. But it is important because many of the objections to the identity theory rest on confusions about what it means to say that a mental state and a brain state have all the same properties.

We have already encountered one such objection in our discussion of dualism. You will recall that one of the initial attractions of dualism stems from the fact that we seem to know all sorts of things about the states of our minds without at the same time knowing things about the relevant states of our bodies. This would seem to imply that minds and bodies must be different sorts of thing, but as we showed with the help of the example of lightning being nothing more than an electrical discharge, the appearance is deceptive. People have come to learn that a flash of lightning is the same thing as a (particular sort of) discharge of electricity. Similarly, so the identity theorist argues, we are now learning, with the help of sciences such as neurophysiology and biophysics, that a mental state is the same thing as a physical state. To be more precise, each mental state of a person is identical with (is the same thing as) a particular physical state. Although this would entail that the mental state and the physical state had all the same properties, the identity theory is not thereby shown to be false, as the dualist claims it is: the theory escapes because to say that someone knows, say, that he has a pain, is not to describe a property of the pain; so it will be no objection to the identity theory that such a person does *not* know that his brain is in a certain state.

Another objection to the identity theory arises from not being clear exactly what it is that the identity theorist claims to be identical with what. Consider the following 'thought experiment'. Suppose a neurosurgeon were to expose your brain while you were conscious. (Lest this sound too gruesome, we would ensure that you had a local anaesthetic for the operation; furthermore, any probing the surgeon did into your brain wouldn't hurt, since the brain is without feeling.) You stare at a bright yellow light which is hidden from the view of the surgeon, who is examining your brain. One obvious way for the surgeon to discover what conscious experience you are having is to ask you. But could he find out just by inspecting your brain? We might be tempted to say that he obviously couldn't and that therefore the identity theory is false. Of course a surgeon couldn't actually do this now, but this might be just because neurophysiology is still a comparatively new and developing science. However, the objector believes that no matter how far it develops it would be impossible for someone to describe someone else's conscious experience just by inspecting his brain-tissue. When I gaze at a yellow light, my experience is of something which is yellow. But if my experience of seeing something yellow is identical with a physical process in my brain, then aren't we forced to say that something in my brain is yellow (and any theory which forces us to say this must be wrong, since the tissue of the brain is pinkish)? This is, however, no argument against the identity theory, because the identity theorist doesn't say that an object of perception (the yellow thing I see) is identical with a piece of brain tissue; he says, in this case, that it is the *experience* of seeing something yellow which is identical with a physical event in the brain. And experiences don't have any colour.

There are indeed other objections to the identity theory which focus attention on what exactly it means to say that a mental state is identical with a physical state, and that a mental event is identical with a physical event. We are not going to discuss any more of them here. However, we do hope to have shown you the sort of way an identity theorist would try to rebut them.

The identity theory and rationality You will recall that another of the apparent attractions of dualism is that it seems to provide an account of human freedom which is compatible with there being a mechanistic physical world. We argued that this attempt is not successful, and that explaining human freedom is one of the most difficult problems in philosophy. Notwithstanding this, the identity theory does, on the face of it, seem less able than dualism to explain the possibility of rational human action. So it is important, if one is to defend the identity theory, to be able to give some account of human action which is compatible with the theory. What sort of account could an identity theorist give?

If the identity theorist is to have any conception of rational action at all it must be one which can be made to square with the principle of universal causality: the mind is made up of physical stuff, and it and its workings must be bound by causal laws like anything else. If I have the belief that grass is green, or the desire to fly in Concorde, then my belief or desire must be thought of as physical states of my body (more particularly, of the brain and central nervous system). As physical states they have prior causes of an equally physical kind; there are no breaks in the chain of cause and effect, which we could attribute to the workings of an immaterial mind.

If I do something, such as get on a train to Wolverhampton, and am asked why I did what I did, I do not reply that I was caused to take this train by physical states and processes in the brain. Indeed I have only the vaguest idea of what physical states and processes do reside in my brain. What I might say in answer to the question is that I wanted, say, to watch a football match in Wolverhampton.

In order to explain my action I refer to my desire to watch a football match, and in general, when we explain our projects we talk of beliefs and desires, not of physical processes in the brain. The vocabulary we use in our everyday explanations of each other's actions is *not* the language of biochemistry or physiology. Nevertheless, the identity theorist must hold that when you say that someone believes or desires something, you could express this fact about them in the idioms of some special science such as biochemistry. The identity theorist must acknowledge that this is at least possible since he holds that beliefs and desires, and other mental states and events, are physical.

Suppose that I did catch a train because I wanted to see a football match. What is this want? According to the identity theorist, it is a state of the brain. As such it has causes, which in turn have further causes, and so on back to a time before I even existed. I obviously did not choose these causes which were responsible for my having a particular desire at a certain time. So how could I be said to be free? I caught the train because I wanted to, but I did not choose to have the want I in fact had.

Belief is similar. It is not up to me to believe what I like. If I see it is raining then I cannot choose to believe that the sun is shining, no matter how much I try. Beliefs as well as desires are caused and are not subject to one's choice.

What these considerations show is that the identity theory does rule out a certain sort of freedom, which many people think they have. But they don't tell against the possibility of people making rational decisions, having rational beliefs, and acting rationally. If, for instance, someone tells me that a belief of mine is wrong I am able to think about the reasons he gives and about the reasons that I have for holding the belief. As a result of this reflection I may come to abandon my original belief and acquire a new one. Similarly I am open to persuasion about what I do. Reflection can cause me to change beliefs, desires and intentions. It is the ability to consider and be swayed by good arguments which forms the basis of our rationality and, some would say, our humanity. But why should one think that this is incompatible with causation, or the identity theory? If I change a belief in the light of a good argument I am *caused* to re-evaluate my belief by what is said to me, and as a result of this I cancel my belief in the light of the new considerations. The cancellation of my original belief depends not upon my free choice, but on the quality of the argument. The distinctively human activity of reflection then, does not seem to be incompatible with causation. And if this conclusion is right the identity theorist oughtn't to be accused of downgrading humanity by denying it the possibility of rationality.

Concluding remarks We cannot pretend to have outlined all of the objections which have been levelled against the identity theory by philosophers, though we have tried to provide a fair sample. The identity theory accords well with recent scientific findings, and it is the theory to which the authors incline – but you shouldn't accept it on that account alone! We end the chapter with this thought: if the identity theory is true, then the death of the body will be the death of the person, and certain religious views about disembodied life after death – views which square naturally with dualism – will be ruled out. So the theory has important consequences for our views not only in the rest of philosophy, but beyond philosophy as well.

3

Philosophy of science

WHAT IS PHILOSOPHY OF SCIENCE?

What is a scientific explanation? What is a law of nature? How can we tell that one theory is better than another? These are typical of the questions philosophers of science seek to answer.

The philosophy of science concerns itself with the analysis of the structure, concepts, methods and results of science. The questions it tries to answer fall, roughly speaking, into two classes. The first class includes questions about concepts which belong to specific sciences, such as 'mass', 'velocity' and 'time' in physics. The second class of questions is about the general logic or methods of scientific practice, and this class includes the questions we began with above. We are going to be concerned in this chapter mainly with philosophical descriptions of scientific practice in general, and little will be said about the analysis of concepts which belong to the specific sciences; discussion of them would presuppose a fair knowledge of those sciences themselves. Here we do not presuppose such knowledge.

Although we shall not assume a scientific background on the part of the reader, we shall need to refer to examples of concrete scientific practice from time to time – otherwise we should have nothing to philosophise about! The reader who feels that the whole of science and mathematics is a totally closed book is urged to read through at least one of the books listed at the end of the chapter.

However, even those readers who feel they know nothing about science are probably more influenced by its outlook and methods than they know. To see this consider Hesiod's account of the creation of the universe:

Verily first of all did Chaos come into being and then broad-bosomed Gaia [earth], a firm seat of all things for ever, and misty Tartaros in a recess of broad-wayed earth, and Eros, who is fairest among immortal Gods, looser of limbs, and subdues in their breasts the mind and thoughtful counsel of all gods and all men. Out of Chaos, Erebos and black Night came into being; and from Night, again, came Aither and Day, whom she conceived and bore after mingling in love with Erebos. And Earth first of all brought forth starry Ouranos [sky], equal to herself, to cover her completely round about, to be a firm seat for the blessed Gods for ever. Then she brought forth tall Mountains, lovely haunts of the divine Nymphs who dwell in woody Mountains. (Hesiod, *Theogony*, lines 116 ff., trans. Kirk and Raven (1957))

This passage is a mythical account of the origin of the world composed at some time during the seventh century B.C. Why do we call it 'mythical' rather than 'scientific'? We shall attempt later to list some of the conditions which scientific explanations have to satisfy.

One more example will be helpful. Consider the following two arguments designed to refute Galileo's theory that the earth moves round the sun.

1 Animals which move, have limbs and muscles; the earth has no limbs or muscles, therefore it does not move.
2 The theory of the earth's motion is against the nature of the earth itself, because the earth is not only cold but contains in itself the principle of cold; but cold is opposed to motion and even destroys it – as is evident in animals, which become motionless when they become cold.

These arguments were put forward by Polacco in 1644 under the auspices of the Church just after the condemnation of Galileo. Do you consider them to be scientific arguments? If you do not, then you *do* have a conception of scientific practice even though you might disclaim any knowledge of science! Part of the task of the general logic or methodology of scientific practice is to attempt to reveal what this conception is and then think about the reasons for accepting or rejecting it. (As we shall see later, we shall find, as happens so often in philosophy, that we have to deal with a *multiplicity* of conflicting ideas about the nature of scientific practice.)

Why should philosophers think about science at all? Surely, it might be argued, the only people in a position to do this are practising scientists themselves (this point has been made on numberless occasions by a chemist to one of the authors of this book). The fact that we have chosen to write something on the philosophy of science shows that we reject this argument, and it is important to see why we do.

We said earlier that the analysis of concepts which belong to specific sciences does require knowledge of those sciences; and here, quite plainly, scientists themselves are in as good a position, and often a better one, to undertake such analysis. Einstein once remarked that the physicist knows best where the shoe pinches in physical theory. Einstein himself was forced to examine certain ideas in classical physics which had been taken for granted by scientists since the seventeenth century, such as the ideas scientists had about the nature of space and time.

But this kind of self-conscious reflection on concepts employed by scientists (as opposed to the use of these concepts in their everyday work) typically occurs only when something seems to be seriously wrong. Ordinarily the working scientist takes for granted a collection of mathematical and physical concepts which he uses in the course of his investigations; he does not reflect on these concepts themselves. A physicist involved in quantum mechanics will, more often than not, be concerned with details about the properties of some sub-atomic particles, rather than with the fundamental principles of the theory. Botanists are likely to be concerned with the ways in which new plants might fit into an existing classification rather than with the overthrow of the whole system of classification which is currently accepted. Nor will his activity as a classifier be likely to be the subject of much of his attention – this is taken for granted by him, and needs to be if he is going to get started doing botany at all.

What we have termed 'general logic or methodology of science' seeks to extract from the specific sciences a set of general principles which describe the nature of scientific theorising. Scientists themselves are often too enmeshed within a specific area of science to do this; and many lack the inclination. This suggests that the philosopher of science can inquire into the nature of scientific practice from a more general perspective than that of a specific scientific discipline. But this leads to another question, namely, what is the point of philosophical inquiries into science?

In the first place we live in an age when the significance of morality and choice is increasingly being called into question. Some of the reasons for this are due to the results and implications, or alleged results and implications, of science. This view is often encapsulated in slogans like 'Science shows us that human freedom is an illusion', and 'Man is only a machine'. Slogans like this are apt to generate a good deal of heat and unhelpful rhetoric. What does it mean to say that science has shown the illusory nature of human freedom? Precisely which scientific results are supposed to support the claim that man is just a machine? The philosophy of science can often be useful for reflecting upon such claims and their significance for other areas of human concern.

Secondly, philosophers have always been interested in concepts like 'truth', 'justification' and 'knowledge'. The human species has been claimed to be unique on this planet in its ability to behave and think rationally. We all have beliefs for which we can give reasons, such reasons being an indication of the truth of these beliefs. The importance of having true rather than false beliefs is obvious in every sphere of life – imagine being operated on by a doctor who believed that your heart is where your brain is, and vice versa! Since the seventeenth century science has developed a series of theories of great generality and predictive power. A natural hope is that there are reasons which underlie their success, and if we can discover what they are, they ought to help us with the traditional great questions of philosophy like 'How can we achieve knowledge and avoid error?' It is worth dwelling upon this point since the holding of incorrect beliefs can be fatal to our ability to control and order our environment in the way in which we would like.

Many of us would not be impressed by someone who told us that convulsions were a sign of the victim's being possessed by demons, as some once believed. We prefer the view that convulsions are caused by electric discharges in the brain; and that consequently the victim is to be pitied and treated rather than feared and shunned. We may say that given the evidence that supports the electric discharge theory, it would not be rational to support the demon view. But this invites the question: What is it that makes it rational to favour the electric discharge view rather than the demon view? Consider the following dialogue:

A I believe that invisible demons cause convulsions.

B But that belief is irrational and hence absurd.

A What do you mean by calling it 'irrational'?

B I mean that we now know that it is unscientific to believe in demons, and that there is a better explanation to hand, namely that convulsions are caused by electrical discharges in the brain.

A All you mean by calling the demon theory unscientific is that scientists choose to ignore it and prefer an alternative theory. 'Scientific' is just a word which applies to whatever it is that scientists want to investigate. Since they prefer to investigate electrical discharges in the brain (which incidentally are themselves caused by the demons) rather than demons, they call the belief in demons 'unscientific'.

It is a good idea at this stage to think through for yourself why you would side with either A or B, or partly with both, in this little dialogue. (Later, you will be able to reconsider it in the light of our discussion of scientific explanation.) But, even if you should decide that it is impossible to take the demon hypothesis seriously, it is not implausible to suppose that you might hold some belief or other which many other people find impossible to take seriously. We take a current example which has divided people sharply both inside and outside the scientific community. Consider the claims of Uri Geller; he claims amongst other feats to be able to bend forks and spoons by the power of the mind. The Geller controversy provides a striking contemporary instance of conflict of belief.

In November 1973 Geller appeared on a B.B.C. television show with John Taylor, a mathematical physicist at the University of London. Geller performed his by now familiar feat of bending, or appearing to bend, a spoon by mental effort. Geller has convinced thousands of people that he is able to bend metal objects by mental action. On the other hand many others, scientists and non-scientists, are convinced that Geller is a fraud who simply uses a conjuring trick to produce the illusion of psychic power. For present purposes it does not matter whether you think that Geller is a psychic or a fraud; two books are listed at the end of the chapter which would help you weigh up the pros and cons. What does matter is that this is a question which, in essentials, duplicates many of the features of the demon argument, and it is not settled by simple observation; we should see exactly the same things happening whether Geller's claims are true or not. In fact it is surprising how much more there is to science than just seeing things; for instance, we see nothing in the everyday world that assures us of the truth of Newton's first law of motion (that bodies not acted upon by a force would coast along for ever in a straight line in uniform motion). On the contrary, our everyday sensory experience would seem to suggest that bodies not acted upon by a force come to rest!

Taylor in his book *Superminds* has this to say about Geller's feats of metal bending: 'One clear observation of Geller in action had an overwhelming effect on me. I felt as if the whole framework with which I viewed the world had suddenly been destroyed' (Taylor 1976, p. 56). Taylor has gone on to try to fit the 'Geller phenomenon' into a scientific framework. This, he says, is necessary in order to 'preserve the scientific understanding of life as the legitimate one'. He acknowledges that many of his scientific colleagues refuse to accept the reality of psychic events, and says that he finds this understandable on account of their determination to cling to a particular framework of theory.

In the journal *New Scientist* for 17 October 1974, Dr Joseph Hanlon presented a lengthy report on investigations carried out by scientists with Geller's cooperation in the United States, and on the results of the *New Scientist* inquiry into Geller's alleged psychic powers. Hanlon, disagreeing with Taylor, concluded that Geller is simply a magician who has fooled scientists who are not familiar with the tricks of professional conjurers. On 28 November a letter appeared in the letter columns of the *New Scientist* which read in part:

Sir – I write to express my sincere sympathy to you and the scientific community concerning the very real difficulty in which you find yourselves over Uri Geller.

As you perceive so clearly, admission of any of the claims made by or on behalf of Geller would knock the bottom out of your science and knock the bottom out of many of you!

In the light of these conflicting beliefs we can construct a dialogue which strongly resembles the demon dialogue:

A I believe that Geller has a psychic power which enables him to bend metal objects.

B Actually, to say he bends metal objects by means of psychic power is not an explanation, it is to stake the claim that an explanation is called for, but we won't go into that. The fact is that we already have a perfectly adequate explanation for Geller's metal-bending feats, namely that he is a magician who by means of a conjuring trick produces the illusion of bending metal objects by non-physical means. Nothing in contemporary science could account for the phenomenon if we take it at face value.

A That is my point! Because you cannot account for it in terms of contemporary science you prefer the explanation that Geller is just an illusionist who pretends to have psychic power. In other words, if science cannot explain something it does not exist. And since Geller maintains that psychic phenomena do exist and he won't accept that he accomplishes his feats by conjuring tricks he must be a fraud. Why won't you consider instead the possibility that Geller is telling the truth and that it is contemporary science which is inadequate?

B But science has developed ways of deciding whether things which are claimed to exist really do. We know, for instance, that a perpetual motion machine is impossible because its existence would violate the laws of thermodynamics. So even if I should see a person 'demonstrate' a perpetual motion machine, I would know that there must be a hidden source of energy.

A Here we go again! Perhaps the man who demonstrated this machine is not a fraud; perhaps there is something wrong with your precious laws of thermodynamics. After all, scientific laws are not laws promulgated by scientists telling nature how to behave, they are statements recording what scientists think they have found out. But the universe is a big place and the observations of scientists are restricted to a very small region of it. You must learn to keep an open mind!

How are we to make a start at resolving the Geller question? In general, what sort of considerations ought we to bear in mind when contemplating clashes of belief about the way the world is? Notice that, in the dialogue on Geller, B appealed to the idea of explanation and to the idea of the acceptability of theories. What are scientific explanations and just when are theories adequate? We are going to leave the Geller dispute up in the air at this stage, since the rest of the chapter is devoted to some of the important ideas that we need to be clear about as a prelude to making a serious assessment of this and other cases where our beliefs about the nature of reality conflict.

SCIENTIFIC EXPLANATION

Here is a list of some of the things that scientists are often said to be doing:

Deductively drawing conclusions from hypotheses. (Newton showed deductively that the proposition that the planets sweep out equal areas in equal times as they go round the sun follows from the proposition that there is a force directed towards the sun.)
Explaining events or classes of events. (The kinetic theory of gases explains why the temperature of a gas goes up as the gas is compressed.)
Confirming theories by observation and experiment. (The observation that light is refracted (bent) when it enters glass confirms the wave theory of light, which predicts that this will happen.)
Disconfirming (or disproving) theories by observation and experiment. (The first voyage around the world disconfirmed the theory that the earth was flat.)
Generalising from observation and experiment. (Galileo generalised from his experiments with balls on inclined planes to the behaviour of all bodies in motion.)

It is, clearly, not only scientists who do all of these things. Each of us in our daily lives continually acts in ways which we could classify under these headings, although we don't usually think about our actions in this way. The child sees a stone sink in water and forms the judgement that all stones sink in water (*generalisation*); we know that either Mary or John went to the party, and then we learn that John got sick and stayed in bed, so we infer that Mary went (*deduction*); we *confirm* our belief that the door is locked by trying to open it and failing; we *explain* the failure of our car to start by citing the fact that the sparking plugs are wet. In these and countless other ways we behave as scientists behave, although often our behaviour is quite unreflective. A lot of people suppose that there must be a great divide between scientific thinking and the way we all think in our daily lives as a matter of course, because scientists have developed special technical ideas and vocabularies for their investigations of nature. Both we and the physicist observe that bodies fall to the ground, but whereas we might content ourselves with the simple statement 'All bodies fall to the ground', the physicist would say 'Whenever an object has been released from altitude for free fall *in vacuo* for a period of t seconds it will move towards the earth's surface with a velocity of $\frac{1}{2} gt^2$.' For ordinary purposes the former statement is quite sufficient to help us steer a safe route around the world. If I jump from the top of the Post Office Tower I know what will happen. But the bald statement that bodies fall to the ground is obviously inadequate when we get curious enough to ask questions like 'Do heavy bodies fall faster than light bodies?' or 'Does the speed of a falling body depend upon the distance it has travelled or the time that has elapsed since it started falling?' These more refined questions call for more elaborate answers like the one quoted above, but – and this is the important point – when scientists give them they are only doing what each of us does already at a less sophisticated level. So when we think philosophically about the ideas listed at the beginning of this section we are doing something more than trying to get clear about how science works; we are attempting to understand the principles which underlie our capacity to learn anything at all.

Let us start with the concept of 'deduction'. 'Deductive' is a term which applies to arguments, and an argument is a series of statements each of which is either a premise or a conclusion. The statements called conclusions are those which are deduced from the statements called premises. The crucial thing about deductive arguments is that they are either valid or invalid. If they are valid, then it is impossible for the premises to be true and for the conclusion(s) to be false. Notice that we call an *argument* valid or invalid, and we call the *statements* which make up the argument true or false.

This is a bit abstract, but quite deliberately so, since it is very general. We encounter deductive arguments in every area of science (including mathematics, which consists entirely of deductive arguments) and in everyday life. The statements which figure in deductive arguments can be expressed either in English, or in the language of a specific science such as physics or geometry. When a scientist deduces a certain consequence from his theory we can think of the theory as a list of statements (the premises) and the deduced consequence as the conclusion. Similarly, when the Euclidean geometer proves a theorem from the axioms of Euclid's geometry, the axioms are the premises and the theorem is the conclusion.

Let's look at an example of a simple valid deductive argument:

1 If that sample of water is heated to 100 degrees centigrade then it will boil.
2 That sample of water has been heated to 100 degrees centigrade.
3 Therefore that sample of water will boil.

Here, statements 1 and 2 are the premises and statement 3 is the conclusion. The argument is deductively valid because it is impossible for the premises to be true and the conclusion false. Notice that this is not at all the same as saying that the premises *are* true. Consider this next argument which has exactly the same form as the one just given:

1 If that sample of water is heated to 100 degrees centigrade then it will freeze.
2 That sample of water has been heated to 100 degrees centigrade.
3 Therefore that sample of water will freeze.

Now we know that it is false that water freezes when it is boiled, so premise 1 is unacceptable. But the argument remains valid, because if premise 1 *were* true, and if a certain sample of water were heated to 100 degree centigrade, then it *would* be true to say that the sample of water would freeze.

If we think about both of the arguments given, what we notice is that they have the same shape or form, which is this:

1 If A then B.
2 A.
3 Therefore B.

('A' stands for 'That sample of water is heated to 100 degrees centigrade' and 'B' stands for either 'That sample of water will boil' or 'That sample of water will freeze'.) We can put any true or false statements we like in place of 'A' and 'B' and the resulting argument will still be valid. This is so, to repeat, because to say that an argument is valid is not to say that the premises or conclusion *are* true, but only that *if* the premises were true then the conclusion would also have to be true. This means of course that even where we have established that an argument is valid we still don't know if the conclusion is true; we still have to find that out by discovering whether the premises are true. Suppose, for instance, a scientist offers us some conclusion which follows deductively from a certain hypothesis. Then what we know, in advance of any observation or experiment, is that if the hypothesis is true then the conclusion must be true; we don't yet know if the hypothesis is actually true.

Many people encountering this account of validity for the first time suppose that deduction can't be very useful if it does not tell us without further enquiry whether the statements we use as premises (or our theories, which are sets of statements) are true or false. The apparent paradox is that the great merit of deductive argument is that it can proceed without our knowing whether our premises are true or false; that is to say it is possible to identify valid reasoning in science and in everyday life without considering truth and falsity. The branch of knowledge which seeks to identify valid argument patterns like the one just considered is called logic. The argument patterns identified by logic are used in, and control, mathematical and scientific reasoning. Why should deductive argument be of such great value?

Many scientific theories cannot be directly tested. This means that we need deductive reasoning in order to deduce from the theories consequences that *can* be tested. For an example, consider Newton's first law of motion, 'Every body free of impressed forces either perseveres in a state of rest or in uniform rectilinear motion *ad infinitum*.' This law – the law of inertia – is not self-evident in the sense that we encounter anything in our everyday experience which assures us of its truth. Nor is it evident to 'the natural light of reason'. Indeed, a long line of thinkers from Aristotle to Galileo have disputed the truth of the law. Think, for instance, of the tugging or the pushing of a cart. This seems to suggest that a steady force is required to keep objects in motion; and that in the absence of a force they would come to rest. So, how can we learn whether to believe the law? What is meant by saying that bodies continue in motion to infinity? Further, since according to Newton every particle of matter in the universe exerts a force on every other particle, we will never be in a position to make an observation on a force-free body. How, therefore, can we tell whether bodies would behave in the absence of forces as the law says they would behave?

Although the law is not directly testable against our observations, if we link it with other physical information the law does have consequences which are testable; we can deduce, using mathematics and logic, descriptions of how the world will behave in certain respects that we can observe (we cannot observe bodies coasting along for an infinite time). Newton, for instance, deduces from his three laws of motion and certain other assumptions that in one second the moon, in order to stay in its orbit round the earth, deviates by 0.0536 inches every second from the path it would follow if there were no forces acting upon it. And this result is in excellent agreement with observation (of course, the observations involve more than merely gazing in the direction of the moon on a cloudless night!). So deductive reasoning enables scientists to put their theories to (often unexpected) tests, though it is important that the finding of deductive consequences of theories which are in agreement with observation does not make it inevitable that these theories are true; we return to this crucial point in the next section.

Having clarified what is involved in the idea of a deductive argument we now turn to the important notion of scientific explanation. Explanations are usually called for when something occurs which puzzles us – something 'out of the ordinary' or surprising. Good scientists are often those who find everyday events puzzling, as the story about Newton and the apple illustrates. Why does water turn to steam? Why do unsupported bodies fall to the earth? Why do children resemble their parents? These are typical of the questions which science seeks to answer.

Consider, for example, the explanation which science gives in answer to the question why water turns to steam. Water is composed of tiny bits of matter (molecules) which are in constant motion. At normal temperatures, the molecules which compose the water are held together by forces operating between them. But if the temperature of the water is increased beyond a certain point the molecules acquire sufficient energy to overcome the forces which hold them together. So they fly apart and escape into the atmosphere. This is what happens when water is heated and gives off steam. What is it about this account that makes it an explanation? According to the view which we are going to sketch, explanation essentially involves deduction, although a successful explanation requires more than just this. We can illustrate this by considering some 'explanations' which are unsatisfactory.

Suppose someone were to say that the reason why water turns to steam is that water is an odourless liquid. We would not think that water being odourless could explain water turning to steam when heated because we can see no connection between the first fact and the second. But what does 'connection' mean here? According to our view the relevant connection is a deductive one; the phenomenon to be explained (or rather the statement which describes the phenomenon to be explained) must be deducible from the statements which constitute the explanation. Therefore the statement

1 Water is an odourless liquid.

does not explain why it is true that

2 Water turns to steam when heated.

because (2) does not deductively follow from (1). Proof: the argument has the form

1 A (= Water is an odourless liquid).
2 Therefore B (= Water turns to steam when heated).

But this argument form is invalid as the following instance of it shows:

1 A (= Water is an odourless liquid).
2 Therefore B (= Water is inflammable).

Since it is true that water is an odourless liquid and false that water is inflammable the original argument was invalid. Recollect that if an argument is valid then if the premises are true the conclusion must also be true.

It might occur to you to say that although citing the fact that water is an odourless liquid is not by itself an explanation of why water turns to steam when heated, we can get an explanation if we add something more to what we have already. Suppose as part of the explanation we add the statement that all odourless liquids turn to steam when heated. We then have the following 'explanation':

1 Water is an odourless liquid. 2 All odourless liquids turn to steam when heated.	} Statements which constitute the explanation
3 *Therefore* Water turns to steam when heated.	} Phenomenon to be explained

Now this argument *is* valid; it is not possible that the statements which, taken together, constitute the explanation be true and the conclusion false, so (3) is a deductive consequence of (1) and (2). Do we now have an explanation for why water turns to steam? No, because it is false that all odourless liquids turn to steam when heated, and we cannot explain what is true, namely that water turns to steam when heated, by appealing to what is false, namely that all odourless liquids turn to steam when heated. So we must say that, although the above 'explanation' satisfies the condition that the statement describing the phenomenon to be explained is a deductive consequence of the statements which constitute the explanation, we still don't have a satisfactory explanation because part of it is false. It therefore appears that we must impose a second constraint upon scientific explanations, namely that they must be true.

This may *seem* a very obvious requirement to impose upon explanations, but it's not at all obvious that the claim is even plausible. It could be objected that we will never know when our explanations are true, and as a consequence we will never be able to say of any such purported explanation that it satisfies the requirement of truth. The history of science is full of examples of theories which were held to be true, and later discarded. If we learn anything from the history of science we learn that we should be wary of claiming that we *know* that our present theories are true. If we do not know that our present theories are true, then we do not know whether they provide scientific explanations or not, so it looks as if we have to say that we will never be entitled to claim that anything has been explained – which is an implausible conclusion to draw from the imposition of what looked like an obvious constraint.

But the requirement that explanations have to be true is different from the requirement that they have to be *known* to be true, and we shall see below, and in the next section, the difficulties of judging the truth or falsity of scientific theories. In the meantime we can solve our present dilemma (whether to accept an implausible conclusion or to remove a plausible constraint) in one of two ways. We can reduce the severity of the requirement that explanations be true by relaxing our standards. Perhaps theories do not have to be true to be explanations, they just must not be known to be false. Alternatively, we might reject the suggestion that we don't know when some explanations are true. We've been seduced into accepting this by noticing that the history of science is full of examples of theories which were once held to be true but are no longer. But this fact does not support the conclusion that we don't know *any* true explanations *now*; the most it shows is that we don't know that we know any true explanations. This means that as long as we have every reason to believe that an explanation is true, and none to believe it false, then we are justified in saying that it is an explanation.

Suppose then, that we do decide that 'explanations' must be true in order to be genuine explanations. Do we then have all the requirements that must be satisfied by a scientific explanation? We do not. To see why, let's look at an example which Molière justly made famous. Suppose that a certain drug causes all who take it to fall asleep shortly thereafter. We want to know the explanation for this, and someone offers us the explanation that the drug causes people to fall asleep because it has a 'dormitive power'. Since 'dormitive' is an adjective meaning 'causing sleep', this explanation of why the drug causes sleep amounts to saying that the drug causes sleep because it causes sleep. The 'explanation' doesn't give us any information about the world. It says nothing about the course of our actual or potential experience which we did not know already. We started out from the fact that the drug had a certain effect; and our 'explanation' simply redescribes what we already know, without going beyond it.

Stated as baldly as this, the 'empirical content' constaint (that all explanations must have consequences for our potential experience beyond the experience of the effect that is being explained) seems natural and obvious; too obvious to be worth mentioning. But consider this 'explanation': 'Bodies fall to the ground because they are under the influence of gravity.' This isn't an explanation as it stands, yet many people, if asked why bodies fall to the ground, respond by offering something like it as an explanation – you can check this for yourself. It is not an explanation of why bodies fall to the ground if your sole reason for believing that gravity is in the offing is that bodies fall to the ground! In this case all you mean by 'gravity' is something like 'force which makes bodies fall to the ground'! Of course it is true that physicists do make reference to gravitation; but such references are a kind of shorthand way of referring to a law of nature together with a background theory. And such references amount to far more than merely a simple verbal redescription of prescientific observation.

There is, then, an intuitive plausibility in the empirical content constraint on explanations, but we may wonder whether there isn't something more fundamental underlying it. Consider a person who maintains that the reason water boils is that it is being affected by the actions of an invisible weightless spirit. Is it without empirical content to maintain that steaming water contains a water spirit? Surely this claim involves more than merely a redescription of heated water?

In order to bring out what is important in this apparently silly example let us look at another example. Consider how the concept of 'caloric' was used in physics. In order to explain why, for example, a hot piece of iron dropped in water raises the temperature of the water, it was postulated that there exists a fluid, called 'caloric', whose sole properties are heat and the capacity to transfer between bodies in contact. Like other fluids, it flows from higher to lower concentrations, and these can be measured by the thermometer. So if we place a piece of hot iron in cool water we find that eventually the iron and the water have the same temperature, which is greater than the initial temperature of the water and lower than the initial temperature of the iron; that is, the iron and the water come to have the same concentration of caloric. Now let us ask: does the statement 'Thermometers measure the concentration of caloric in bodies' have empirical content?

Caloric (like our spirit) was held to be an invisible weightless fluid. (It had to be weightless since it was found that a piece of iron did not get heavier when it was heated.) So is not the concept of 'caloric' without empirical content, since someone who claims that this fluid exists would not be saying that one should expect, as a result, to have some particular set of experiences, rather than another? If you want to say that it is not, because thermometers actually measure caloric concentration, why should we not reply with equal plausibility that in fact what thermometers measure is not concentrations of caloric but the activities of invisible and weightless spirits?!

Notice that although the idea of caloric as a weightless fluid was not easy to fit into the rest of our picture of the physical world, it was, for a time, a very useful idea. For instance, we can measure out caloric in units. The amount of caloric which will raise the temperature of one gram of water by one degree is called a calorie. Also, the picture of heat as a substance which flows from one body to another just as water flows from one level to another appears to fit many of the facts of heat exchange. For example, just as the total amount of water is conserved when we pour some from one bottle to another, so the total amount of heat is conserved when heat flows from one body to another.

Does this mean, then, that the theory of heat as a *substance* – caloric – ('the substance theory') has empirical content? What we should notice is that the theory broke down because it was limited to problems of heat flow. It could not explain, for instance, heat generated by friction. So what condemned the theory was not the charge that it lacked empirical content but the charge that there were heat phenomena which the theory ought to have been able to explain but could not. But now that we have a better theory in the shape of the kinetic theory of heat, which can account for many facts which the substance theory had no explanation for, we might be inclined to say: really the substance theory lacked empirical content all along. For a time it looked as if it might have it, but we know now it did not.

But this claim ignores a vital difference between the spirit theory and the substance theory which it is important to bring out, and which talk of empirical content tends to obscure. Suppose we have three facts, A, B and C. (We can suppose that they are facts concerning the phenomenon of heat such as the fact that heat flows from hot bodies to cold bodies but not vice versa.) Now imagine that we have three explanations E1, E2 and E3. Suppose further that E1 explains fact A but fails to explain facts B and C, E2 explains facts A and B but not fact C, and E3 explains facts A, B and C. If other things are equal we would obviously prefer the explanation E3 to either E2 or E1, and the explanation E2 to E1. Why? The reason is that E3 explains more than E2 or E1. We prefer explanations which have the widest scope. And what underlies this preference is our desire to understand as much of the world as possible. We can equate increase in the generality of our explanations with increase in scientific understanding. Scientists value highly theories which reveal systematic interconnections between the widest possible range of hitherto unconnected facts.

These considerations bear upon the spirit and caloric theories in the following way: although (in the light of the modern kinetic theory of heat) we wish to discard both the spirit and the substance theories, they do not appear to be unsatisfactory because they are purely verbal or circular. These, you recall, were the grounds on which we rejected the 'explanation' that a certain drug induces sleep because of its dormitive power. Intuitively both the spirit and the substance theories have more content than this. Also, we want to be able to distinguish between the spirit theory and the substance theory with respect to their empirical content. So what the empirical content constraint on explanations requires is that a set of statements may only constitute an explanation if the number of unexplained facts is reduced by it, since this results in an increase in our understanding of the world. What facts does the spirit explanation promise to connect up? According to it, water turns to steam because of the action of invisible and weightless spirits. What else does it explain? Nothing. It points to no connection between this fact and any other fact connected with heat. We gain nothing by offering it as an explanation. In fact, suppose instead we said that water turns to steam when heated because heating water causes the spirits to vacate the water; and spiritless water always steams! Why is this any worse a thing to say than the former? In neither case do we advance beyond the fact that water turns to steam when heated. The spirit theory stands condemned because it fails to increase our understanding of the world. And it fails to do this because no reduction in the number of apparently disconnected facts about heat is brought about by postulating it. By contrast, the substance theory, although it was eventually overturned by the kinetic theory, did connect together disparate facts about heat which need explanation.

What underlies the empirical content constraint, then, is a principle of economy of explanatory principles – the more facts of an apparently disconnected sort that a theory can explain, the more powerful it is as an explanation, and the more empirical content it has. The example we considered which seemed to flout the constraint was one in which there was only one type of fact which was being explained, which makes its empirical content about nil. Scientists consider explanations in sciences like mechanics so powerful just because the same set of concepts (such as mass, velocity and time) have explained such seemingly diverse facts as the shape of planetary orbits, the motions of the tides, and why a bicycle pump gets very hot if the pump is rapidly compressed with the outlet partially sealed!

We have now considered three requirements that acceptable explanations must satisfy; the deductive connection requirement, the truth requirement, and the empirical content constraint. Is there anything further that we need to add? Yes, there is the further important requirement that any explanation must contain at least one *universal law*. A universal law is a statement which asserts that everything of such-and-such a kind has so-and-so properties. Universal laws hold with unrestricted generality, which means that they purport to say something which is true everywhere in the universe. Let us examine two illustrations of universal laws.

First, consider the fact that light travels at 300,000 kilometres per second. The constancy of the velocity of light is a universal law which can be expressed like this:

All propagation of light is such that $s = 300{,}000 \times t$ where s is the space traversed in kilometres and t is the time taken in seconds.

The law refers to all samples of light, past, present and future, and to all regions of space. Next consider the ideal gas law. This law states that the pressure of a gas (P) multiplied by its volume (V) is equal to the temperature (T) of the gas multiplied by a constant (R). We can bring out its universal form by writing the law thus:

> Every sample of gas behaves in a way such that its pressure, volume, and temperature are in the relation PV = RT.

The two laws just looked at, and scientific laws in general, have the form of *universal, conditional* statements. Why are scientific laws conditional and universal?

Let us first of all consider the conditionality of scientific laws. A conditional statement has the form 'If A then B', where 'A' and 'B' are statements. A conditional statement does not assert the truth of the statements conditionally linked. Consider the conditional statement 'If this gas is compressed it will increase in temperature.' This statement neither asserts that the gas is compressed nor that the gas increases in temperature; what is asserted is that *if* the gas is compressed *then* it will increase in temperature. Suppose that all gas were to disappear from the universe tomorrow (we ignore the objection that if this were to happen much else would also disappear). Would we then want to say that the gas law was false – because there happened to be nothing for it to apply to? Intuitively, it seems reasonable to think that were all gas to disappear tomorrow then the gas law would still hold, on the grounds that if gas reappeared then the law would hold of it. But there is a deeper reason for requiring that scientific laws be conditional in form, a reason which goes beyond the desire to accommodate science-fiction cases.

Consider again Newton's first law of motion (see page 101). We noted that the law says what happens to bodies not acted upon by forces; and that in fact all bodies in the universe are acted upon by forces. The power of the law resides in the fact that although it makes reference to what happens in situations which, as a matter of fact, never occur, when the law is coupled with other statements we can provide powerful explanations for many observable phenomena. Plainly, if the law of inertia asserted the existence of force-free bodies, it would be false. But if the law is conditional, as follows:

If a body is not acted upon by a force then it either perseveres in a state of rest or in uniform rectilinear motion *ad infinitum.*

then we don't have to account it false. For in conditional form the law does not assert the existence of force-free bodies. The process of idealisation and abstraction from what is given in sense-perception is a central feature of human mentality, strikingly exemplified in scientific theorising; we, unlike other animals, are able to entertain hypothetical or conditional statements. A scientist will often reason, 'Such-and-such is taking place. But if so-and-so factors were not present then such-and-such would not occur.' And in science it is often the case that the unfulfilled conditional expresses the deeper insight than the fulfilled one. Compare the fulfilled conditional 'If you drop this one pound lead weight and this feather from 10 metres above the ground then the one pound lead weight will reach the ground first' with the unfulfilled conditional 'If there is no air around the earth then a one pound lead weight and a feather dropped from 10 metres will reach the ground together.' We know that the unfulfilled conditional has greater generality of application.

Next, consider the universality of laws; the fact that they make reference to *all* rather than to some, or one. The reason laws must be universal is that it is only in this way that we can satisfy the requirement that the statement describing what is to be explained is a deductive consequence of the statements which constitute the explanation. To see this consider the following 'explanation' of why a certain body after free-falling from rest for 2 seconds has traversed an interval of 32 feet:

1 The distance a freely falling body falls from rest is sometimes equal to $\frac{1}{2} \times 16 \times t^2$.
2 This body had been falling from rest for 2 seconds.
3 Therefore this body has fallen $\frac{1}{2} \times 16 \times t^2 = 32$ feet.

Here, statement (1) is the law; (2) expresses what are called the *initial conditions*; and (3) is the deduction made from (1) and (2), which is in agreement with observation. The only trouble with this is that (3) is not a deductive consequence of statements

(1) and (2). This means that we have not explained why the body in question has traversed an interval of 32 feet after freely falling from rest for 2 seconds. The law statement (1) only says what happens to *some* falling bodies, and how do we know that the falling body in question is one of those? We are only justified in deducing (3) from (1) and (2) if we replace 'sometimes' in (1) by 'always'. Thus amended, (1) expresses Galileo's law of falling bodies, and we do indeed have an explanation for the event to be explained. Thus, laws must be universal in form if conclusions are to be validly drawn from explanations (laws together with statements of initial conditions).

The explanation just discussed is a typical case of explanation in science. Notice that our account makes explanation and prediction two sides of the same coin: if we start by knowing the law of falling bodies (statement (1) in the above example with 'always' instead of 'sometimes') and the initial conditions (2), then we can *predict* what will happen; we will know in advance of observation that (3) is true. If on the other hand we begin by observing the event described by statement (3), we can *explain* it by citing (1) and (2).

While this account of explanation seems to characterise one important class of explanations in science satisfactorily, it has to be stressed that quite a few contemporary philosophers of science believe it does not adequately describe statistical explanations. We may cite the statistical generalisation that such-and-such a percentage of people who smoke *x* cigarettes for *y* years develop lung cancer, in order to explain the death from lung cancer of Mr Smith, who has smoked *x* cigarettes for *y* years. This kind of explanation violates the deductive consequence rule for explanations because the generalisation cited to explain the death of Mr Smith does not rule out that Mr Jones, who has also smoked *x* cigarettes for *y* years, will *not* contract lung cancer. Some defenders of the deductive model of explanation are prepared to argue that, for this reason, statistical generalisations are not really explanations at all. At best they are a sign that something needs explanation ('we shall not explain lung cancer until we can pin down the precise factors that cause it'). The philosophy of statistical explanation is a complex and difficult topic and we shall not discuss it further here.

We have said, then, that a central pattern of explanation in science is this: from a universal law or laws, together with statements describing initial conditions, is deduced a statement describing the event or type of event to be explained. Notice, however, that we have *not* argued that all explanations which satisfy the conditions we have discussed are satisfactory; only that explanations which fail to satisfy them are unsatisfactory.

We have been arguing for the following position:

> Science provides explanations for why things happen. Scientific explanations increase our understanding of the world. (They do this, we recall, by revealing connections between apparently unconnected facts.)

Now consider the following counter-claim:

> Science does not explain why things happen; it merely tells us how they happen. Therefore scientific explanations don't really enable us to understand the world (although they may enable us to control part of it).

Spelled out in a little more detail, this counter-claim runs as follows: If we ask a scientist *how* objects behave when they are released from a given altitude above the earth, he tells us that they move towards the earth's surface with a velocity of $\frac{1}{2}gt^2$. By contrast, if we ask him *why* falling bodies should behave in this way, his only answer is that they just do. If we press him further he may derive the law of falling bodies from another general statement (in science it is not uncommon for one generalisation to get explained by another) but sooner or later he just has to stop and admit that his explanations (or series of explanations) involve general principles which cannot themselves be explained in terms of anything else. The most general expression of the worry that underlies the counter-claim is that science is not competent to deal with the really deep questions about the nature of the universe which come to us all in moments of profound reflection. Why does the universe exist? Why is the law of inertia true? Science, it may be argued, is powerless to answer these questions since science has to assume certain general principles and laws in order to get the process of explanation off the ground at all; but since these general principles and laws are themselves part of the universe, they also need to be explained.

This is not a new line of thought. Richard Bentley asked Newton the question 'Why is there one body in our solar system qualified to give light to all the rest?' and Newton replied that he knew of no reason but 'Because the author of the system thought it convenient'. One might argue that Newton's reply to Bentley was unsatisfactory because one could then go on to ask 'Why did the author of the system think it convenient?' And if the reply to this further question is that he just did, we might as well rest content that, similarly, it just is the case that there is only one sun in our solar system.

Could science ever explain why the universe exists? And if the answer to this question is negative should we conclude that scientific enquiry is somehow inadequate or limited? We shall argue that science could not provide an explanation for the existence of the universe, but that it does not follow from this fact that scientific explanations are thereby made inadequate or limited. Consider the following three questions:

1 Why is the law of inertia true?
2 Why do planets go round the sun in elliptical orbits?
3 Why does the universe exist?

At present, science cannot explain why the law of inertia is true; the law has no known cause. It just seems to be a basic feature of the universe that bodies continue in a state of rest or uniform motion unless acted upon by a force. It is possible that one day science will provide an explanation for the law of inertia by deducing it from some other principle or principles; the law is not essentially one which is incapable of being explained in terms of some other principles, as far as we know. Notice that it does not follow from the claim that science cannot explain everything, that there is one particular thing (such as the law of inertia) that science *could* not explain, any more than it follows from the fact that I cannot count all the natural numbers (1,2,3,4 etc.) that there is one particular number that I could not count up to.

Now consider question (2): science can explain why the planets go round the sun in ellipses. We can deduce the shape of planetary orbits from the statement that there is a force directed towards the sun, together with some mathematics. Notice that the explanation for one feature of the universe (that the planets go round the sun in ellipses) is given in terms of another (that there is a force directed towards the sun). In general the explanations we give for features of the universe themselves make reference to features of the universe. How about question (3)? If by 'universe' we mean by definition everything that there is, we obviously cannot find an explanation for the existence of the universe. This is so because according to our model of explanation we should need a set of statements including at least one universal law and a statement describing some initial conditions. But the features described in the law and initial conditions would themselves be *part* of the universe, and hence part of what we were supposed to be explaining. Our view then is not that scientific explanation is somehow limited because of its inability to answer question (3), but that there is nothing that *could* count as an answer to question (3).

When people hear these sort of reflections for the first time there is a tendency for them to suppose that this is some kind of cheat, an attempt to exercise a scientific imperialism. 'You say that if science cannot explain why the universe exists then nothing can explain it. But why should we accept this?' Perhaps for answers to these very deep questions we need to appeal to mysticism or religion. We do not have the space to explore fully this line of thought, but we will say the following. If we took the line that we need the hypothesis of a Judaeo-Christian Creator (for instance) in order to explain the existence of the universe, then either we have to say that there must be some other explanation for His existence or we have to say that He explains His own existence. Even supposing that we could make sense of the idea that a thing could explain its own existence, further argument would be required to explain why the universe was not such a thing. In short, to say that God is the explanation for, the cause of, the universe is to replace one question with another which is at least as puzzling, namely 'What is the cause of God?'

It may help a bit if we reflect upon the fact that we seem driven to ask grand questions like (3) just because we can ask of any event or type of event E, 'What is the cause of E?' So we feel we ought to be able to ask 'What is the cause of the universe?' But we should realise that questions which are perfectly proper when asked about individual things and types of thing are not necessarily intelligible when asked about collections or sets of things. Thus, I can ask how tall a certain man is, but it is absurd to ask how tall the human race is. The human race is not the kind of collection which itself has a height. Again, I can ask where a piece fits into the jigsaw, but I must not expect an answer if I ask where the jigsaw fits in. Therefore, we cannot suppose that there is an answer to question (3) just because it makes sense to demand explanations for events and processes going on within the universe.

The reader may still feel dissatisfied with these reflections; but the authors no more expect to obtain an answer to question (3) than they expect to obtain an answer to the question about the height of the human race. Scientific explanation cannot be held to be limited or inadequate because it cannot answer unanswerable questions! We admit, however, that this is a topic which needs much more argument than we have space to give to it.

Is it possible that one day science will succeed in explaining all happenings in the world in terms of a small set of laws? In the nineteenth century, this was certainly a widespread hope, but nowadays scientists are less optimistic. Some think that the search for the fundamental laws of sub-atomic physics (for example) is like the activity of peeling layers off a kind of infinitely-layered onion: as soon as one layer is uncovered, the next stands ready to be removed. Others think that the search for deeper and deeper explanations will eventually cease because scientific research will become more and more costly. It would be foolish for us to take a stand on this question here.

CONFIRMATION OF LAWS AND THEORIES

We have been emphasising that one of the roles that science plays in our life is to provide well-tested explanations of everyday occurrences. In this section we are going to leave aside considerations of what constitutes an explanation and concentrate on the ways in which such explanations are tested. For many people it is this aspect of science which constitutes its rationale, its very reason for existing. When a conflict of belief arises (Jill believes women are equal in intelligence to men, Jack doesn't), an appeal to 'scientific method' is often made in the hope that this procedure will resolve the disagreement in the most rational manner. Before we can decide whether science deserves this kind of respect we will need to be clear about its method. Our description of this method will be double-edged; we won't rest content with just describing what scientists *do* because to some extent our exercise is prescriptive: it will involve us in saying what scientists *ought* to be doing.

A simple, and persuasive, conception of scientific method is embodied in the story of Galileo's dispute with the Church. On the one hand we have the picture of Galileo producing laws concerning the behaviour of falling bodies by the method of experiment and observation; on the other that of the Church dignitaries 'refuting' Galileo by recourse to theological dogma. According to this story, the difference between scientists and non-scientists is that the former will always support their conclusion by appealing to experimental and observational facts; the latter will dogmatically uphold their own views whilst ignoring any relevant evidence which would prove them wrong.

Galileo is seen as the man of science who dropped cannonballs from the Leaning Tower of Pisa to prove that bodies of different weight would fall at the same speed, and who looked through his telescope to discover that Jupiter had moons circling round it, contrary to the opinion of the theologians of his time. The theologians, for their part, are accused of stupidity for ignoring the testimony of their senses, and for even not allowing their senses to provide them with any evidence which might run counter to their beliefs (as in the story that they refused to look through Galileo's telescope).

The truth of this story is not our present concern. What it does is to insinuate a commonly held belief about the distinction between scientific and non-scientific approaches to our beliefs. Science is taken to be predominantly concerned with observation and experimentation. To be scientific is to make observations, do experiments, draw conclusions from the results of these, deduce further consequences from these conclusions for further testing, and so on. Implicit in the reliance on this procedure is the assumption that it is only our experience of the world which will enable us to gain any knowledge about it. Our beliefs are to be proven correct or incorrect by reference to such experience, and any other way of justifying them is irrational. Experimentation, which provides us with tailor-made experiences, is of particular importance. In an experiment we can vary the conditions under which a particular effect has been observed, and this, it is assumed, will give us new knowledge of the different effects such variations will produce. Again, it is assumed that in this manner science will accumulate, via experimentation, more and more knowledge of the universe and its workings.

On this view all conflicts of belief, such as the one concerning the intelligence of women, are easily resolvable. All that needs to be done is for an experiment to be performed, the outcome of which accords with one of the beliefs and contradicts the other. Any other method, such as resorting to 'what the Bible says', is beside the point; and refusing to accept the judgement delivered by the experiment is the height of irrationality.

Is this a satisfactory account of the essentials of scientific method? In order to answer this question we need to make our provisional account clearer on a number of points. We have talked about the scientist as embarking initially on a fact-finding mission, and then drawing conclusions from the facts he has found. Must this be the order of enquiry? Must the experiment precede the formulation of a hypothesis? As soon as the question is formulated in this way a satisfactory answer appears impossible. If we feel that facts must precede hypothesising we are committed to explaining how the selection of facts occurs – in accordance with what principle do we decide that some observations would be relevant to our problem, whilst others would not be? In the case of I.Q. testing we would take into account the age of the tested subjects while ignoring the colour of the pencil with which they wrote their answers. This suggests that there are criteria (in the form of assumptions or hypotheses) which we use to disentangle the important from the irrelevant facts. On the other hand, it appears that we can only form hypotheses with a background of some knowledge – our questions about variability of I.Q. only make sense if we know that there are some facts which we can take for granted (e.g., at a fairly primitive level, that there are people in the world at all).

Our question appears to lead us into a dead-end. Which answer we prefer may not be terribly important: if a new discovery is made in science it doesn't seem very significant whether or not the hypothesis preceded the experimentation. What is crucial is the relation which exists between hypothesis and evidence. We have seen that as far as *explanation* is concerned, the relationship that exists between the hypothesis (or law) and what it is explaining is a deductive relation. However, we are now concerned with the support which the evidence gives to a hypothesis, and a little reflection will show that this cannot consist in a deductive relation.

A deductive argument, we said, is characterised by the impossibility of validly drawing a false conclusion from true premises. When we talk of the support given to a hypothesis by the evidence, the sentences describing the evidence ('Object no 1, weighing x lbs, fell y yards in z seconds', 'Object no 2 . . .', etc.) form the premises which warrant our holding certain conclusions ('Objects of different weight fall at the same speed'). The premises are, typically, the recorded results of observations and experiments, the conclusions being general statements akin to the universal laws that figure in explanations. The number of facts that we know about (observational and experimental results) is only small compared with the number of facts covered by the general law. It is, therefore, quite possible for our premises to be true, while the conclusion is false. That is, the relation of support that exists between evidence (as premises) and hypothesis (as conclusion) is an *inductive*, and not deductive, relation.

What holds for the simple generalisations which we have been using as examples holds also for the more deeply theoretical aspects of science. We seem to be stuck with the dilemma of remaining at the level of reporting the results of particular experiments, or else going beyond the evidence (e.g. in seeking to explain why this experiment turned out in the way in which it did) and taking the risk that our conclusion might be false. We have already seen that the more general an explanation – the more facts it covers – the more we value it. An explanation which sticks as closely to the evidence as possible is not going to have the desired generality, and will probably be superseded by a more powerful explanation which has the potential to explain more facts. There appears, then, to be a tension between the explanatory and confirmatory aims of science. *The more explanatory the hypothesis is, the greater is the risk of falsehood.*

It is tempting to think that one can get rid of this risk by producing more and more evidence in favour of (or perhaps eventually against) the hypothesis in question. The greater the amount of evidence which we know to be in accordance with a theory, the more likely, we believe, the theory is to be true. However, no matter how much favourable evidence is collected, the problem will remain – a false conclusion remains a possibility. The question of the rationality of believing a hypothesis to be true remains: at what stage have we assembled *sufficient* evidence in its favour for us to be *justified* in accepting the hypothesis and moving on to another problem?

This problem has baffled philosophers for many centuries, and isn't going to be solved within the compass of these pages. It is known as 'the problem of induction'. There is one feature of the evidence, however, which may alleviate some of the apparent tension between the desire for explanatory power and the desire for *reliable* (i.e. unlikely-to-be-false) explanations. So far we have talked of confirmation as though all that is involved is the quantity of evidence in favour of a hypothesis. In the simple cases we have been talking about this *is* all that is involved, and this does lead to a tension with explanation. Let us take three simple generalisations that we might want to test, in order to determine what our beliefs should be.

A All students are red-headed.
B All students in England are red-headed.
C All students at the University of Bradford are red-headed.

Suppose that we are restricted in our evidence by having sent out questionnaires to all the students at the University of Bradford, and to nobody else, and the results show that all these students are red-headed. Which of the three generalisations should we adopt? If we were to say that the best-confirmed hypothesis is the most acceptable, then we would be driven to adopt C at the expense of the more general B and A. C, in fact, is verified –we have exhausted the possible evidence and have not come across a disconfirming instance – i.e. a case which is not in conformity with the generalisation, a case which would here be a non-red-headed University of Bradford student. In terms of *content*, hypothesis A is to be preferred – it contains more information and would therefore, on our account of explanation, have more explanatory power than B or C.

In these cases of simple generalisation, then, it does appear that we have conflicting criteria determining our choice of hypothesis. However, the very simplicity of the cases might be leading us astray. Firstly, the case represented above is not one which is typical of hypothesis-choice in science. Note that our evidence accords with all three hypotheses, and none of the hypotheses conflicts with the other two. Usually when one is faced with choosing between rival explanations in science the hypotheses are incompatible with one another, which enables the experimenter to devise experiments the outcome of which, he hopes, will be in accordance with the prediction of one hypothesis while disconfirming the others – we will return to this point later. Secondly, the nature of the evidence in the above example is of a fairly uniform kind – students who are red-headed. Because of this uniformity, quantity is all that matters; but consider the case where the temperature of a gas is considered to be dependent upon its pressure and volume. Testing this hypothesis will involve varying the pressure and volume of gases to determine the effect, if any, of this variation upon temperature. The *variety* of evidence can be as important as its quantity in the support given to a hypothesis, and the greater the scope of the supposed explanation, the greater is the variety of evidence which can be assembled to support it. If we have an explanation of electricity and a (different) explanation of magnetism, then our desire for economy of explanatory principles (i.e. for more powerful explanations) will make a unified explanation of electromagnetic phenomena preferable – and the very diversity of the phenomena thus explained will increase the credibility of the explanation.

Increased explanatory power, then, allows for a greater variety of evidence to be adduced in favour of the explanation. There is another connection between the nature of explanation and confirmation. We suggested above that sometimes it arises that a test or experiment is supposed to help us decide which of two competing hypotheses is to be preferred. Obviously, such a procedure is only possible if the rival hypotheses are genuinely conflicting, and in this case that means that they will predict different outcomes of the experiment. The severity of the test – its importance as a means of deciding between the hypotheses – is going to depend upon the precision with which the predictions are made. The significance of exactitude and precision can be seen in Kepler's struggle to describe correctly the orbit of Mars. He started off with the Copernican view that the planets moved around the sun in a circle, but could not make this view accord with the evidence available to him. Copernicus had been satisfied with an inaccuracy of up to 0·17 degrees; the accuracy of the new evidence which was available after Copernicus meant that such a degree of inaccuracy was no longer tolerable. Kepler's initial hypothesis contained an error of 0·13 degrees, which was only rectified by the adoption of an alternative view – that the orbit was elliptical.

This exemplifies the two sides of the role that exactness and accuracy play in such tests or observations. The one is that the circle and ellipse hypotheses were formulated precisely enough to give significantly different predictions; the second is that the observations made were themselves sufficiently accurate to render such precision meaningful. An example of how such exactitude of prediction can be rendered useless would arise if two rival theories of belief-formation predicted different results. One such theory (A) could be that we believe what we do as a result of our position in the class-structure of our society, and in particular as a result of the particular job which is allocated to us within that structure. The second theory (B) might explain our beliefs as being dependent upon childhood experiences, in particular sexual experiences. These rival theories might be formulated with a great deal of precision, employing a scale whereby intensity of belief is measured much as a thermometer measures temperature. In order to ascertain which is the better explanation we have to produce a case for which they give different predictions. Say we produce just such a case – a man who, on theory A, should believe that he is being persecuted by the state, such belief being of intensity 36°. Theory B likewise predicts the existence of the belief about being persecuted, but predicts that the strength of the belief is 55°. The two theories are sufficiently precise to give conflicting predictions, but unless we have an accurate and exact way of measuring differences in 'strengths of belief', such precision is unhelpful.

The acceptability of an explanation, then, depends at least upon the amount, diversity, and precision of evidence which supports it. However, it is clear that we do not have any mechanical method which we can apply in science to determine whether any hypothesis is, in fact, true. What we do have are a number of 'rules of thumb' which enable a scientist to make a considered judgement in cases which are being disputed. Some of these rules would emerge from our characterisation of the evidence which best supports a hypothesis – e.g., in cases of dispute, increase the precision of the measuring instruments as much as possible, etc. There is another such rule which hasn't been touched upon yet because we've been looking at the problem of 'confirming' hypotheses as though it was concerned only with the relation between a hypothesis and the evidence which supports (or fails to support) it. An example of the 'refutation' of an explanation by an experiment should highlight the role played by background knowledge and alternative explanation in our acceptance or rejection of hypotheses.

Until towards the end of the eighteenth century the process of combustion (burning in air) and calcination (the formation of metal oxides, which are a kind of ash) had been explained by the so-called 'phlogiston theory'. Briefly, this theory stated that when different ores were heated to form metals a 'metalising principle' was absorbed by the ores. This principle was called 'phlogiston', a substance which was responsible for the formation of metals when their 'earths' were heated. This substance was ultimately related to fire; substances rich in phlogiston would catch fire quickly. It was thought that charcoal was such a substance, which, on being heated, gave off phlogiston, which was then taken in by the ore being heated, to form a metal. When the metal was then burned it was said to give off phlogiston to the air, and to become a calx (so, for instance, iron would become iron oxide). Air which had been used in combustion became heavily 'phlogisticated', and phlogisticated air would no longer support either combustion or animal breathing. Of course we now know that it is not the presence of phlogiston which inhibits breathing, but the absence of oxygen.

The phlogiston theory was formulated at the beginning of the eighteenth century by Becher and Stahl. A consequence of the theory, one would assume, is that since a metal gives off a substance (phlogiston) in the process of calcination, the subsequent calx should weigh less than the original metal. However this 'prediction' turned out to be false; a *gain* in weight was observed. What is of importance to us is that this result was known for decades before the phlogiston theory was eventually rejected as being inadequate. Why did it take such a long time for the 'refutation' to take effect? One of the reasons was, simply, that the theory was useful despite this unfortunate consequence; there were other areas of combustion and respiration where it seemed to function quite adequately as an explanation. It was only seriously threatened when an alternative explanation of these facts was produced, the new hypothesis (developed by Lavoisier between 1770 and 1780) being that oxygen was taken in by a metal when it became a calx. The new hypothesis could explain the same facts as the phlogiston theory – and in addition explained the gain in weight of a metal after burning: since the metal took in oxygen, its weight would increase.

It was only with development of such a rival theory that the phlogiston theory was seriously threatened by the experimental counter-evidence. Even then we do not find that all of its proponents were prepared to give it up in favour of the oxygen hypothesis. Some of its more fervent advocates took the anomalous experimental conclusion to be an indication that the theory required just a bit of 'patching up', rather than outright rejection. A variety of proposals was offered which had the effect of accounting for the gain in weight of a substance after it had been burnt – but most of these had the effect of complicating an explanation which previously had the virtue of being a reasonably simple hypothesis. One of these modifications is of interest to us because it illustrates the importance of background beliefs (i.e. beliefs which do not appear to be directly questioned by the specific test which has upset the theory) in the process of theory change. The phlogiston theory predicted a decrease in the weight of a metal during combustion, or on being heated, this prediction being dependent upon the background belief that the substance being given off in the process (phlogiston) had positive weight. If this assumption is replaced by the belief that phlogiston has 'negative weight', then the resulting theory would correctly predict that when a metal is heated it gains in weight.

Such adjustments to a theory are always possible to make; as a result, the process of theory choice is even more complicated than it originally appeared. At first it seemed that in order to choose between two or more conflicting theories all we would have to do is to devise an experiment whose outcome would be predicted by one of the theories and not by the others. We have now seen that not even this procedure is decisive – the defenders of the 'incorrect' explanation can always locate the problem in one of the background beliefs, rather than in the explanation itself. The reason for this is that the hypotheses being tested are never totally independent of the rest of our scientific beliefs, and it is this interdependence which enables us to locate errors in the linked beliefs rather than in the hypothesis which seems to be being directly tested. This doesn't nullify the effect of counter-evidence: what it does do is suggest that whenever such counter-evidence is discovered it is a matter of choice as to which of the assumptions responsible for the incorrect prediction should be given up. A 'rule of thumb' to be followed in making such a choice is suggested by the following examples.

In 1950 Immanuel Velikovsky published a book entitled *Worlds in Collision.* According to Velikovsky it was the stopping of the earth's spin which caused the Red Sea to divide when, according to the Bible, Moses stretched out his hand, permitting the children of Israel to cross in safety before the waves engulfed the pursuing Egyptians. Velikovsky's theory is that the apparent harmony of our solar system is a comparatively recent phenomenon; that in past times the orbits of the planets intersected, causing collisions between planets which brought about the existence of comets. At the time of Moses (about 1500 B.C.) one of these comets nearly collided with the earth, which twice passed through its tail. The proximity of the comet to the earth caused the earth to stop spinning or to slow down. Eventually the comet collided with Mars, lost its tail and became a planet itself – the planet Venus. Few scientists take Velikovsky's theory seriously.

According to certain people called *astrologers* the position of the planets and stars at the moment of a person's birth determines certain facts about that person's future life, such as how many children he will have, his life span, and so on. The vast majority of scientists find belief in astrology absurd.

Some people believe in various kinds of paranormal phenomena such as the ability to read minds and the capacity to act on physical things by using mental effort only. There are scientists who take such claims seriously, but the vast majority do not.

It is often alleged that the failure of scientists to take the foregoing kinds of claim seriously is due to prejudice or to a perverse unwillingness to face facts which do not fit in with the preconceptions of the scientific establishment. In the light of our previous discussion, such preconceptions would be the set of background beliefs, assumed knowledge, which the scientists are holding constant (i.e. not modifying or rejecting) in order to discriminate between new claims.

One such alleged preconception often cited is that scientists believe in 'materialism' and that this doctrine seriously inhibits the investigation of some kinds of knowledge claims – in particular, the claims of astrology and parapsychology. What is this doctrine of materialism supposed to be? There is probably no one thing which people have in mind when they talk about 'materialism'. In the eighteenth century many philosopher-scientists, followers of Descartes, rejected Newton's theory of gravitation because it seemed to them that Newton was committed to 'occultism', to a belief in a mysterious, intangible force, *gravity*, which reached out from the sun and influenced the path of the earth. How could one thing influence another across empty space? Modern science has incorporated the concept of the field; and no one today, scientist or otherwise, supposes that scientific discussions of gravitation and electromagnetism are occult. As our scientific conception of nature changes, so does our conception of what is real and what is occult. This line of thought leads us to the conclusion that to say science is materialist is just a way of saying that science deals with whatever science deals with. By definition the objects of scientific investigation are material. If on the other hand the claim that science is materialistic is supposed to mean that science does not deal with immaterial 'things' then it begs the question in favour of there being such things.

Let us now consider Velikovsky's claim. On the surface of it it looks scientific and materialist enough! Why shouldn't it be entertained as a serious theory? Quite simply, what is wrong with this theory (at least) is that there is nothing in contemporary physics which suggests that a large body coming close to the earth *would* slow down or stop the rotation of our planet. In fact mechanics says this would not happen. So it looks as if the defender of this theory must reject a considerable body of scientific theory and experimentation. The cost of accepting Velikovsky's theory, then, would be that we would have to give up an enormous number of background assumptions which we had taken to be well-confirmed by scientific experiment.

How about the claims of astrology? The problem with these claims is that how the stars and planets have the claimed effect is unspecified. Given this lack of specificity, and the fact that there is nothing in modern physics which would incline us to believe that such a relationship between the stars and human lives exists, we are justified in allowing the burden of proof to fall on the shoulders of the astrologers. Some astrologers have attempted to make their claims more precise, suggesting that gravitational forces between the planets and stars and ourselves (which do exist) might have astrological effects. No one can prove they don't, yet if they did it would be rather odd, since the gravitational forces produced by the doctor and nurse at birth are far greater than those produced by the fixed stars and planets.

The case of the existence of paranormal phenomena, like Geller's 'powers', again has to be assessed against our current theories in science; we have to make judgements of credibility relative to what we think we know already about the world.

Naturally these arguments depend for their force upon acceptance of contemporary science, and we admit that no one can ever *prove* that our current theories are true, but it does not follow from that that we should not make our judgement on the basis of the theories we do possess. We have to start somewhere.

All of this may sound just like conservative prejudice, but it is in fact a reasonable step to take. The conservative principle at work is, in a nutshell, *when confronting phenomena for which you have no explanation, or rival explanations, accept that explanation which makes as few modifications in your existing beliefs about the world as is necessary.* We see this principle at work in the minds of those who claim that certain paranormal phenomena (such as psychic spoon-bending) are fraudulent. They hold current scientific beliefs which are incompatible with the existence of any mechanism which could account for such spoon-bending, so they maintain their scientific beliefs and reject these alleged facts.

The justification for this conservatism is that the background beliefs with which we operate, the scientific theories which we take for granted, have been accepted because they have proved themselves to be consistent with a wide variety of evidence; to accept an explanation of an event which contradicts these beliefs is to reject what have been well-confirmed theories, and to do that we need very good reasons. This conservatism has been objected to on the grounds that it stultifies the growth of revolutionary science, innovation being outlawed on methodological grounds. Those who make such objections point to periods in the history of science (e.g. the switch from Newtonian theory to relativity) where it would appear that this methodology would have retarded the progress of science by rejecting the new theory.

Two things need to be said about this objection. Firstly, it is not always the case that the principle of conservatism (a principle of 'minimum mutilation') would lead to the rejection of the new theory. In the phlogiston–oxygen debate, accepting the hypothesis about the negative weight of phlogiston would probably have caused more perturbation in existing beliefs (e.g. about the nature of substances) than accepting the new theory, that oxygen was taken in by the metal. Secondly, the principle is not overriding. Sometimes it is the case that a wholesale revision of beliefs about a particular process is made necessary by new experimental findings. Although the patching-up of existing beliefs can go on indefinitely, there may come a time when the resulting theory is so bizarre and complicated that a new, simpler theory would be preferred. This situation seems to have occurred in the change from a Ptolemaic (geocentric) theory of the universe to a Copernican (heliocentric) theory.

Finally, it can be said that the objectors to the principle of conservatism justify their radicalism by a very selective choice of examples. The history of science is usually written as a succession of success-stories; the vastly greater number of failures is omitted. Given this selection of evidence, it appears more reasonable to support the unorthodoxy of an innovator against the 'dogmatism' of a conservative. It is clear that there is sometimes virtue in going against current opinion; it can, sometimes, lead to startling new discoveries and theories. The sting is in the 'sometimes'. For every genius who successfully replaced existing orthodoxy by a new and surprising theory, there are innumerable others whose unorthodox beliefs proved to be unworkable. When supporters of, e.g., Geller accuse his critics of being hidebound by a scientific dogmatism, they may just be complaining of a justified wariness of extravagant claims.

The appeal is made (as it also is in the arts, on behalf of some new but now critically acclaimed composer or artist) to the fact that we may be making a mistake, that we can only make a historically-bound judgement which subsequent generations may overturn. It is true that we may be wrong; that is the price we pay for making any inductively based judgement at all. The alternative is not to embrace every innovator as a genius – for that could also lead to mistakes being made. The only alternative to our historically-bound judgements is total scepticism, the claim that we don't have any reasons for preferring one belief to another. We prefer the position that endorses some judgements against others, and doesn't prize novelty for novelty's sake.

We ended the first section of this chapter with an imaginary conversation between two people about metal-bending. Go back to that dialogue now, and reconsider it in the light of our discussions in the latter two sections of this chaper. We hope that our discussion will help you to see more clearly what underlies their differing views, and thereby use philosophy of science to reach your own conclusion on the controversy.

4

Philosophy of language

The ability to communicate with one another is something that we usually take for granted. We are constantly exercising this ability in a totally unselfconscious manner, using language in a variety of ways. Stepping back from this activity and reflecting upon it, philosophers become puzzled at our facility with symbols which do not appear to have any direct relationship to the objects of the world that they pick out: the symbols are not caused by the objects, nor do they resemble them. How is it possible for us to use such symbolic systems? How does language differ from other forms of communication? Are humans unique in being the only species with the ability to talk to one another? How are theories of meaning related to other philosophical questions?

These are the questions which will concern us in this chapter. In the first section we will be examining the claim that it is only human beings who possess a linguistic ability and competence. In order to do this we will also have to consider the general question of what marks off a language from other potential communicative devices. In the second section we will turn to the last of the above questions – the relationship between theories of meaning and other areas of philosophy. It has been suggested that twentieth-century philosophy has made a 'linguistic turn', that it is predominantly concerned with the analysis of language and the construction of theories of meaning. In the second section we will indicate some reasons why language has become particularly important.

COULD ANIMALS USE LANGUAGE?

In previous chapters we have alluded to the fact that the human being appears to be a peculiar kind of animal – one that has a mind, is conscious of others and of itself in a way that other animals do not appear to be. It is the unique capacities associated with 'having a mind' that justify us in treating human beings differently from other animals – for example, we do not credit dogs or cats with a 'free will', as agents responsible for their actions, and so the punishment meted out to them can be viewed as an attempt to condition them to behave differently, rather than as moral retribution. Furthermore, most of us think that it is all right to kill animals for food, but quite wrong to do this to human beings. This distinctiveness of human beings was emphasised by the dualists, who attributed a mental substance to human beings alone, regarding other animals as purely physical. The belief that only man has a mental substance was held to be justified by the fact that only humans had a linguistic capacity; other animals, it was believed, did not, and could not, communicate with each other using a language. Descartes, for instance, claimed that a monkey-like machine could be constructed which would be indistinguishable from a monkey, but a machine indistinguishable from a human being could not be built. He held this for two reasons: that the adaptability of human behaviour could not be replicated, and that the animal-automaton could never 'use speech or other signs as we do when placing our thoughts on record for the benefit of others' (Descartes 1637, p. 116).

The lack of language was associated with the lack of any reasoning capacity. Most often our reasoning, arguing, drawing conclusions from premises, relating beliefs to one another, is done with the aid of language – so much so that if we didn't have a language it is doubtful whether we could carry out these abstract thought-processes at all. An example used by Colin Blakemore in the 1976 Reith Lectures illustrates this well. Rafael, a chimpanzee, had been taught to gain access to his food by extinguishing a burning wick with a jar of water. He also knew that he could cool himself off by scooping up river water and throwing it over his head. Another trick Rafael had learnt was to get from one raft to another by erecting a makeshift bridge (a bamboo pole) between the two. Faced with the situation of being on a raft with food at one end and a burning wick blocking his path to the food, Rafael noticed the jar of water on a nearby raft. His solution to the problem was to use his bamboo stick as a bridge, fetching the jar of water in order to douse the flame which was obstructing his path to the food. What he had not noticed was that the river water which he had been using to cool himself would have served just as well and would have cost him less effort. An explanation of this 'failure' could be that he had not identified the substance in the jar as being the same as the substance he used to cool himself – he did not have the concept 'water'. Rafael's generalising ability was restricted. The problem for the investigator is how to account for this. What is cause and what is effect? Is it the lack of language that leads to the inability to generalise, or is it this inability which accounts for the lack of language?

The answer probably lies between the two – a certain amount of generalising is required for a linguistic ability to develop, and this development will, reciprocally, improve the generalising capacity. What is clear is that much of our reasoning is done with the aid of language. The dualists' claim was, in fact, stronger than this – it wasn't, they thought, just such *logical* abilities which required the possession of a language, but the ability to think *at all* depended upon the capacity to express thoughts in a language. Thought-processes are also required by mental activities other than reasoning – intending, believing, hoping, wishing, imagining, and so on. If all of these require thought, and thought requires a linguistic capacity, then we cannot attribute any of these mental acts to an animal which lacks a language.

There are two reasons for pursuing the question of whether animals have, or could have, a language. One is that if the above consequences of being languageless are true, then it is an important matter to settle, as it could affect our treatment of, and attitude to, animals. Much of what has been written about 'animal liberation' suggests that our behaviour towards other animals is immoral and unjust, and the justification for this claim would be that we persistently ignore facts about animals, which, if recognised, would lead us to treat them less brutally. One such fact would be the animals' capacity to experience pain. If we adopt the moral position that our actions should not increase pain, then our belief that animals experience pain would have immediate consequences for our actions towards them. However, if we believe that the capacity to experience pain requires self-consciousness on the part of the subject of pain, and that this self-consciousness requires a thinking capacity which in turn presupposes the ability to use language, then we would not have the same attitude towards languageless animals as we have towards humans.

The second reason is this: it is of general theoretical interest to discover what are the special features of linguistic behaviour since possession of language is one of the most distinctive features of our humanity. One way of setting about such a discovery is to compare and contrast our own obvious linguistic ability with more doubtful cases, as this could make us clearer about what it is that we have and animals are said to lack.

What are the characteristic features of language? What is it to have a language, to communicate with others using a language? An immediate suggestion might be: it is the ability to express our thoughts vocally to others, and to interpret the vocal production of the thoughts of others. This raises more questions than it answers: How are thoughts related to the vocal signs of them? How do I know the code which enables me to interpret other people's noises? Is the same code applicable to everybody? Do people speaking different languages have the same thoughts? Despite the difficulty of these questions (and perhaps because of them) the suggestion is not without merit. It emphasises one crucial factor, and that is the communicative function of language. Using language, we can inform people about the world, ask them questions, command them to do things, aesthetically please them – and, in turn, we can be informed, questioned, commanded, pleased. It is true that we do not always use language in such a communicating manner – soliloquising isn't confined to the stage. But this seems to be a parasitic use of language, dependent upon the more important social function it has.

This communicative aspect of linguistic behaviour is what accounts for something which the dualists were trying to emphasise – our mode of living, which appears so drastically different from that of non-human animals. We have a rich heritage of scientific beliefs to draw upon when faced with problematic situations; if we each had to learn from scratch every truth ever learned, we would not have been able to achieve anything like our present ability to order our environment to suit our needs. We do not learn just from our own experience – we rely a great deal on what others have learned before us and have been able (linguistically) to transmit to us. Newton recognised the importance of this when he said that if he had seen further than others, it was only because he had stood on the shoulders of giants. Language has enabled us to 'store' our experience so that new members of the species do not always start from the same position as previous members. This does not appear to happen in other species – and a likely explanation for this is that we, unlike them, have a language.

So far, then, we have suggested that language is used, typically, to communicate. How does it do this? Or, rather, how do we, using language, succeed in communicating with each other? And is all communicative activity language-dependent? Let's start with the last question. Its answer depends on what exactly we mean by 'communication'.

It is clear that a belief that Jack has can be transmitted to Jill without the use of anything which we would want to call a language. Jack can notice a ship in the distance, and his belief that there is a ship out there can be transmitted to Jill just by his odd behaviour – screwing up his eyes and peering out to sea. Jill will acquire the relevant belief by copying him, thus acquiring a belief which originally only Jack had. In this way a belief of Jack has come to be shared by Jill, and it is shared by her because of

Jack's behaviour. This much can also be done by other animals – the startled flight of one buck can alert the others to the dangerous intruder. No language seems to be involved, for reasons which will become clear. Is communication involved? Well, information has been transmitted – but not all such transmission of information is communication. Jack, by smiling at Jill, may cause her to have the belief that he has a hole in his front tooth. She has acquired new knowledge as a result of Jack's behaviour, but we wouldn't classify this as a case of Jack communicating a belief to Jill, even if Jack knew that he had a hole in his tooth.

Communication appears to involve the *intentional* transmission of beliefs. In none of the above examples is it the case that the 'communicator' *intends* the other person or animal to share his own beliefs. In the first two cases, the behaviour of Jack and the buck would have been the same regardless of whether the other person or animals had been present. In the last example Jack behaved in the way in which he did because Jill was there, but the belief she acquired was not one that he intended her to acquire. If, then, communication is the intentional transmission of beliefs, then our question about whether all communicative activity requires language can be settled if we find a case of such intentional behaviour that isn't linguistic behaviour. Such a case seems to be easily constructed. Take our first example again, except that Jack, instead of just peering out to sea, takes hold of Jill's head and guides it so that she will also notice the ship. Here Jack is intentionally producing in Jill a belief that he wants her to share – but no language is involved in this behaviour. Why not?

One of the crucial features possessed by language which is lacking from Jack's behaviour is its freedom to communicate messages independently of the immediate environment. In the case of the buck, and when Jack moves Jill's head, the message is one that is closely tied to the environment. In the latter case, the knowledge that Jack gets across to Jill is communicated with the help of the state of affairs which the knowledge is knowledge of. Were linguistic behaviour like this we would be restricted like the academics of Lagado in Swift's *Gulliver's Travels*, who communicated by producing objects from a sack on their backs – the objects being those that they wanted to 'talk about'. What is lacking here is the peculiarly symbolic character of language, its freedom from reliance on the situation it can represent. This feature has been used to distinguish linguistic symbols from other signs which can give us information about the world. A typical example of the latter would be the indication of fire by means of smoke. We are able to infer from the existence of smoke that there is a fire – the inference being justified by our knowledge of a causal generalisation linking the two. This causal generalisation could be expressed as: Wherever there is smoke, there is fire. The linguistic symbols which we use are *not* causally related to what they represent in this manner – it is just not true that a certain state of affairs obtains whenever certain linguistic symbols are used. It is our freedom from such direct causal control which enables us to speak about fire when there is no fire present, to inform one another of past experiences, and to predict future occurrences. We are able to speak about what is not in the immediate spatial or temporal environment. (This aspect of our linguistic behaviour is a problem for a behaviourist, who would attempt to explain all behaviour, including linguistic behaviour, as a response to certain stimuli. The apparent freedom from environmental control has the consequence that the stimulus, or set of stimuli, to which linguistic behaviour is meant to be a response is very difficult, if not impossible, to specify.)

The relevance of this discussion to Jack's behaviour is that his ability to communicate to Jill his belief that there was a ship in the distance did depend upon the environment in a way in which linguistic behaviour does not – thus even though he did intentionally transmit a belief of his to another person, he did so without using a language. This isn't to deny that some uses of language are environmentally dependent; what are called 'indexical' expressions – demonstratives (this, that) and pronouns (I, you) – *do* depend upon the context in which they are used to refer to the items that they refer to. It is rather the fact that the *entire* content of Jack's message was conveyed in an essentially environmentally-dependent way that renders his behaviour non-linguistic. In short, language has the potential to be used in a way which 'displaces' it from the place it is describing. (Language can, incidentally, also be used in abstract ways that refer to no particular place, as in the language of mathematics.)

Another aspect of the peculiarly symbolic nature of language is what is sometimes called 'arbitrary denotation'. By this is meant that linguistic symbols are non-iconic – the relationship between such symbols and the world is not that of a picture to what it depicts. Usually there are features of an ordinary picture which resemble the scene depicted – and it is because of the resemblance that we know what it is that is being depicted. The symbols which constitute a linguistic system do not, in general, have any resemblance to what they represent – an exception being onomatopoeic words. The word 'dog' doesn't look (or sound) like a dog. Thus language seems to have the amazing capacity to represent reality without having a straightforward causal or iconic relationship with what it represents, and it is the lack of these features which marks off our previous examples of intentionally communicative behaviour as non-linguistic.

So far we've isolated three conditions that must be met by the transmission of beliefs if it is to be linguistic: the transmission must be intentional; it must be possible for it to be displaced from the situation described; and the symbols used must be non-iconic. These preconditions of linguistic behaviour do throw some doubt on what might otherwise have been thought to be examples of a linguistic ability. Take the first feature, the intentional nature of linguistic behaviour. We have already seen in the case of the buck how information can be conveyed without there being a relevant intention. A more sophisticated example is cited by Jonathan Bennett (Bennett 1976, p. 203). Two dolphins, male and female, were separated in adjacent enclosures; in order for them to receive food the male was required to press one of two paddles, the correct one being indicated by a visual signal presented to the female. The problem for the dolphins was to communicate (from female to male) in order to get food, and the 'communication' was achieved acoustically. Here we appear to have a non-iconic way of intentionally transmitting information. That it is non-iconic seems clear; it is less clearly intentional.

The results could be interpreted in this way: the male dolphin was conditioned to react to acoustic signals from the female dolphin, and so would press the correct paddle simply because he was rewarded with food for doing so. The female dolphin was, likewise, conditioned to respond to a visual stimulus, her response being to make a certain noise, her reward being the food released by the male dolphin's paddle-pushing. This account explains the behaviour of both dolphins without postulating any intention to communicate with one another; it also explains why the female would (as she did) continue to vocalise even when the male was not around – the responses to the visual stimuli continued as long as food was forthcoming.

Was there a causal factor present in this experiment which would eliminate the possibility that the rudiments of linguistic behaviour were present? There doesn't appear to be a natural causal relationship, as in the smoke–fire situation, but the dolphins do appear to have been causally conditioned to respond in a very precise manner to carefully selected aspects of their environment, and this causal conditioning does preclude the kind of flexibility which is associated with our use of language. We could, after the relevant conditioning, say of the female dolphin, 'Whenever a red light was shown she uttered two high-pitched squeaks', and of the male dolphin, 'Whenever he heard two high-pitched squeaks he pressed the left paddle.' These are the kinds of generalisation we associate with the smoke–fire case, and not with linguistic behaviour. It is not the case that, for example, whenever a dog is present the word 'dog' will figure in our conversation, or that whenever a dog is mentioned in conversation a dog is present. The causal conditioning exhibited by the dolphins' behaviour, then, greatly restricts that behaviour, whereas our own linguistic behaviour is not restricted in this way.

Associated with this flexibility is another feature of language which has so far not been discussed. All of the examples so far used have been of 'unstructured units' of behaviour. By this we mean, for example, that the 'meaningful' sounds uttered by the female dolphin were not themselves composed of sounds which were also meaningful. The contrast with a full-blown language is clear – by far the vast majority of our (meaningful) sentences are composed of elements (words) which are themselves also meaningful, and the meaning of the parts contributes to the meaning of the sentences. This is part of what is meant by saying that language has a systematic structure. It is structured because sentences are composed of smaller units. A name ('Peter') contrasts with most sentences in this way – the name is unstructured, not being decomposable into smaller, recombinable units which are themselves meaningful. 'The dog is in the garden' does have a structure, each word being usable in different sentences. The structure is systematic in so far as the precise way in which the words are combined has an effect on the meaning conveyed by the sentences. 'Tom loves Mary' and 'Mary loves Tom' are composed of exactly the same words – only the ordering is different. It is the order in which the words appear which makes the meaning of the two sentences different.

It is this element of language which has fascinated contemporary philosophers and linguists. We are able to construct an enormous number of sentences by learning (*a*) a relatively small number of words, and (*b*) how to combine the words into meaningful sentences. This makes language an extraordinarily powerful instrument of communication, as we are able to say an enormous amount on the basis of mastering a limited vocabulary. This is of interest to both philosophers and linguists because of the dual aspect of the structure of language. When linguists attempt to formulate the rules whereby words are combined into larger *grammatical* units, they are constructing a *syntactic theory* (a theory of the correct way to combine words into sentences – no account being taken of what such sentences mean). Philosophers have, recently, been more concerned with the rules which govern the combination of words into larger *meaningful* units, and in doing this they are doing *semantics* (the science of meaning). Linguists have been struck by a child's ability to produce new *grammatical* sequences to which he has not been exposed; philosophers are intrigued by our ability to understand sentences which we have never heard before. An explanation of this latter ability is that familiar words are combined in an unfamiliar fashion; we understand the words and their rule of combination, thereby understanding new sentences.

Considerations such as this suggest that language has an exceedingly rich structure with great expressive power – and it is, presumably, something with similar power that we would be attributing to other species if we claimed that they communicated using a language. It seems clear, from the evidence available, that Descartes was, in the main, correct – we do not find animals in the wild communicating with each other using anything like such a powerful system as human language. But it is true that the evidence is not conclusive; and that it is *possible* that animals engage in lengthy conversations without our understanding them. Colin Blakemore suggests just that:

> There is no denying, of course, that speech *seems* to outstrip in complexity any animal communication system that we understand. But I wonder if we are not missing immense subtlety in the behaviour of animals. Indeed, it is inconceivable, as Chomsky points out, that monkeys and apes have the capacity for language but have simply never put it to use. Perhaps they do use it, but we cannot comprehend them. (Blakemore 1976, pp. 144–5)

If Blakemore is suggesting merely that it is *possible* for our conclusion (that animals don't have a language) to be wrong, then what he says is correct but remarkably unenlightening. What has to be done is to weigh the evidence at our disposal and accept that conclusion which is most plausible – accepting at the same time the risk that 'perhaps' we're mistaken. Given that we have not yet deciphered the code of animal languages, the evidence suggests that there is no such code to decipher. No evidence that we have will *prove* that animals do not have a language, but we cannot expect such proof for any empirical results. What we know at the moment is that humans do appear to be different from other species in their possession of a linguistic capacity. For some (dualists, for instance) this might indicate a hard and fast distinction between us and other species, which marks a never-to-be-bridged gap separating us off from the rest of nature. However, nothing we have said so far prejudges the issue of whether, although animals may not develop a language in the wild, they could nevertheless be taught a language. If this could not be done then it would suggest that animals are irremediably different from us.

Whether language can be taught to animals is a topical and controversial issue, being the subject of intriguing experiments done with chimpanzees in the last ten years. The most famous of these has concerned Washoe, who has been taught American Sign Language (ASL) since 1966 by Allen and Beatrice Gardner. ASL is considered to have most of the important properties possessed by other human languages, so the success or failure of Washoe's training will be of some significance in determining whether an animal can be taught a language. Given our previous discussion, what will be of interest here is to see whether Washoe has mastered enough of ASL to be able to communicate using new sentences – i.e. signs in different combinations. It will also be important to see whether the rules she has mastered allow for the systematic production of these novel sentences, so that the meanings of the sentences can, at least to some extent, be seen to be a result of the use of the rules.

So far the results of the experiment have been fairly remarkable. At the age of 4 years Washoe was using, in an appropriate fashion, 85 signs, and was producing strings of signs up to 5 signs long. An important result has been the use, by Washoe, of novel sentences – signs in combinations which Washoe had not been taught to use. Washoe also initiated many of the conversations, unlike the subjects of other experiments who were trained by a simple system of directly rewarding the right behaviour. (Washoe's training was more naturalistic – trainers in Washoe's company always used sign language to talk to one another and to Washoe, and she was taught in the way we teach children.)

It is difficult to judge whether or not these results justify the claim that Washoe is using a language to communicate. It is clear that the Gardners have gone much further than previous investigators in teaching a communication system to a non-human animal. The following is a reported dialogue involving Washoe and Dr Roger Fouts.

> Roger picked up an apple and offering it to Washoe said, 'What's this?'
> Washoe seemed to come out of her funk and knuckled over signing, 'Fruit'.
> 'Who fruit?'
> 'Washoe fruit'
> 'What Washoe fruit?'
> 'Please Washoe fruit' (Linden 1975, p. 12f.)

Washoe seems to be able to make her desires known and to communicate fairly efficiently with her human companions. How much more one is prepared to say will depend upon the importance one attaches to further experimental findings. So far the Gardners have analysed 294 two-sign combinations produced by Washoe, the result of such analysis being to place Washoe's performance on the same level as that achieved by children between 16 and 27 months. There is one notable difference reported by Roger Brown (1976), namely that Washoe's combinations have been produced in any order with no apparent change in intended meaning. For children learning their first language this is not the case. That word-order does contribute to differences in meaning has already been remarked upon – reflect upon the difference between 'Washoe tickle' and 'Tickle Washoe'. The same elements are involved in the combinations, but the difference in their order of appearance produces a significant change in meaning.

Does this difference mean that Washoe should not be credited with the ability to use, and communicate with, a language? If such a lack of ordering persists, then it certainly would detract from the claim that Washoe's language has a *systematic* structure, and we have found such a systematic structure to be an important part of the expressive power of our own language. A lot depends upon how one explains the lack of word-order. Roger

Brown, a psychologist, believes that although the presence of particular word-order is evidence of the utterer's intentions that we should understand the sentence in a certain way (as determined by the word-order), the absence of word-order is no clear indication that such intentions are lacking – especially as in nearly all the circumstances in which Washoe uses combinations which, because of lack of word-order, may be ambiguous, the circumstances themselves disambiguate the sentence. For example, if Washoe were playfully to bite a cat, and sign 'Cat bite Washoe', the circumstances, plus the knowledge that for Washoe word-order does not contribute to meaning, would convey the correct message. Brown also suggests that a complicating factor in the experimental set-up is the difference between visual and acoustic codes – Washoe depends upon *seeing* signs, children upon *hearing* words. It is suggested that it might be easier to put *sounds* into a regular order, than to do the same with a group of *visual* signs – thus explaining the otherwise unexplained difference between Washoe and 'stage 1' children (children between 16 and 27 months old, for whom word-order is a crucial feature of communication). It might also just be the case that the experiment has not been going long enough for Washoe to develop the relevant expertise – Blakemore, more recently than Brown, does claim that Washoe has caught on to the difference that word-order makes, but he doesn't cite the source for this claim.

Whether or not we are prepared to admit Washoe to the full-blown 'stage 1' of language learning, it is clear that although much has been achieved more will have to be achieved if we are to be justified in attributing to chimpanzees linguistic behaviour that matches ours in its expressive capacity. Some will see the results already produced as indicating an ability which only requires further training, and improved techniques of training, to be fully realised. Others will see Washoe's ability as still too environmentally controlled, too geared to the needs and wishes of the immediate present, to have that freedom from stimulus control which marks our linguistic behaviour. We have suggested that this feature of language is associated with its symbolic character, the way in which words and sentences acquire a meaning which can be conveyed from one person to another by and large independently of the immediate environment in which the sentence is uttered. In the next part of this chapter we shall turn to a specific example of a theory of meaning, to see why it is that philosophers have found this a particularly fascinating and important area of philosophy.

THEORIES OF MEANING

So far we have looked at one aspect of language which interests philosophers – the way in which the ability to speak a language differentiates us from other animals. In the course of that investigation we noted that an extremely puzzling aspect of language is also its most important property – the phenomenon of the meaningfulness of our discourse. Our ability to talk to one another about almost anything we please is so much a part of our everyday existence that we tend to take it for granted, rarely reflecting upon how it is possible for us to do this. We tend not to remember how it is that we gained our first language, and so our natural view of ourselves is as talking, communicating animals, and something so commonplace and pervasive is easily ignored. When we do reflect upon the nature of language and, in particular, the phenomenon of meaning, it rapidly loses its quality of 'ordinariness' and becomes something extraordinary, requiring some kind of explanation. How is it possible for the noises which come out of our mouths to become meaningful linguistic units? Think of two bilingual people, both of whom know, say, French and German, talking to one another, the one speaking French, the other German. They are using different languages, yet manage to communicate. Is the meaning of the sentences they utter somehow independent of language, so that different languages can be seen to be different embodiments of a more universal, abstract realm of 'meanings'? How do the 'arbitrary signs', which we talked about in the last section, function as meaningful units within a language? What distinguishes non-significant noise (the babbling of a baby) from words which appear to have acquired meaning? These are some of the questions which theories of meaning attempt to answer. This area is of great importance for philosophy in so far as the particular theory of meaning which one adopts might have consequences for one's views in other areas of philosophy – for instance, in social philosophy and philosophy of mind. In this section we shall discuss one such theory and show why its adoption could be influential in these areas. The particular theory discussed is usually called 'verificationism', and its 'verifiability criterion of meaning', as we shall explain, has clear consequences elsewhere in philosophy. Before moving on to that theory, however, it will be illuminating to consider briefly an alternative, probably more 'commonsensical', view of meaning.

People use sentences to say something about the world. This much is uncontroversial, but what is controversial is the view of meaning which it sometimes suggests. According to one theory sentences can only be used to talk about objects in the world if they are related to those objects, and the words that constitute the sentence are meaningful just because they refer to objects in the world. This view takes names as being the prime example of the way language relates to the world. Proper names, such as 'Jack' and 'Jill', are used to pick out individuals who are so named. We understand the name in so far as we understand who it is that is being referred to by somebody using that name. So far, so good. Unfortunately for this theory, not all language consists of proper names, so the extension of the 'naming relation' to other parts of language is problematic. Even as a theory about the meaning of *names* this apparently commonsensical view encounters difficulties. The name 'Jack' can refer to many different people – is its meaning made up of all the people it refers to, or does it change its meaning according to the context in which it is used? The second alternative would appear to be the more plausible, since on the first view every baptism and every death of somebody named 'Jack' would change the meaning of the name. One version of this second alternative switches the focus of investigation from the relationship between words and the world to the relationship between words and the people who use them. A view of meaning along these lines is suggested by John Locke:

> Man had by nature his organs so fashioned as to be fit to frame articulate sounds, which we call words. But this was not enough to produce language. It was further necessary that he should be able to use these sounds as signs of internal conceptions, and to make them stand as marks for the ideas within his own mind, whereby they might be made known to others ... (Locke 1690, p. 223)

Instead of the sounds we utter referring to objects in the external world, and deriving their meaning from that relation, on Locke's view they are signs for ideas within a speaker's mind. But this view renders all communication problematic – how do we know which ideas are relevantly related to another person's utterances? The ideas are 'internal conceptions', and we presumably do not have direct access to, or immediate knowledge of, the inner working of another person's mind. We tend to rely on his behaviour for such knowledge, and the important behaviour in this case is his utterance. The picture suggested by Locke is that we understand the utterances by discovering the internal ideas to which they are related; the problem is that we can only discover the internal ideas by understanding the utterances. So if this picture were correct we would simply not be able to understand one another's utterances.

The verifiability criterion of meaning The verifiability criterion of meaning is a further variant on theories which claim that words derive their meaning from naming things in the world, a variant which attempts to get away from the difficulties of Locke's theory by concentrating on the way in which sentences are discovered to be true or false. The basic idea is that a sentence is meaningful only if there are facts which tell for or against its truth. Sentences which are reasonably straightforward descriptions of the world are held to be unproblematically meaningful, i.e. they seem to be the kind of sentence about whose meaning there is little disagreement: on the whole, people tend to have no trouble communicating with each other using these exemplary sentences. In addition, the truth or falsity of such simple descriptions is fairly easy to assess. These two points are related to one another in the verifiability theory. It is claimed that these sentences (e.g. 'The shirt you are wearing is white') are ones about whose meaning there is little disagreement *because* it's a straightforward matter to discover whether such sentences are true or false (or as philosophers put it, 'to determine their truth-value'). According to the verificationist point of view, one cannot understand what a sentence means without knowing the sort of thing one would have to do to determine its truth-value. A consequence of this view is that sentences whose truth-value just cannot be assessed (such as 'There is a rabbit in this room which is completely imperceptible and doesn't affect the world in any way') are meaningless, or at least lack 'cognitive' (factual) meaning.

An additional motive for holding this view of meaning is the consideration that children learn language by learning from, or imitating, their parents' utterances. These utterances, particularly when aimed at the child, are spoken in 'appropriate' circumstances, and the appropriate circumstances are usually those which the sentences describe. (For example, the one word sentences 'Mummy', 'Daddy', are spoken to the child in the presence of the relevant person.) The child, on this theory of language-learning, acquires a grasp of the meaning of the sentences by relating them to the world in the proper way – and being able to grasp meanings *at all* depends, it is argued, on sentences being able to be related to the world in this way. What, then, could a sentence mean (and how could its meaning be learnt) for which no obviously distinctive 'appropriate circumstances' can be specified?

These considerations are combined with the view that progress in philosophical disputes, such as the ability of philosophers to resolve long-standing disagreement about the nature of 'fundamental reality', have been severely hindered by the fact that the proposals and counterproposals are often couched in a language whose terms do not relate to the observable world in the desired manner. In contrast to the disagreements one finds in science, it appears that this kind of philosophical argument just *cannot* be decided one way or another because there is nothing in the world, or in the universe, which would tell for or against thc views and theories which are being contested. This kind of 'empty' disagreement is thought to be typical of 'metaphysical' philosophy. Idealists, who believe that all that we observe is ultimately constructed out of ideas, or spirit, and Materialists, who believe the basic ingredient to be physical matter, are metaphysicians. The apparent disagreement between their views cannot be settled, it is claimed, because they would agree on all observable facts. The Idealist geographer and the Materialist geographer do not disagree in their respective maps of the world, nor would an Idealist and a Materialist physicist be led by their metaphysical beliefs to hold different physical theories. As their contrasting beliefs cannot be proved right or wrong by citing relevant facts, their views are held to be unverifiable – hence, according to the verifiability criterion of meaning, not meaningful.

It is not our purpose to chart here all the twists and turns taken by the proponents of this theory of meaning in their replies to criticism levelled against it. One charge should be mentioned, however, because of its importance for our attitude towards scientific theorising. The initial formulation of the theory held that a sentence is meaningful only if it is verifiable, with 'verifiable' being taken to mean 'capable of being proved true'. We have already mentioned in our chapter on the philosophy of science that most scientific theories would fail this test – their generality, or universality, entails that we do not have *conclusive proof* about their truth. Put another way, scientific theories cannot be deduced from the sentences describing the facts which are explained by the theory. We may be able to deduce these 'observation-sentences' (in the sense of Chapter 3) from the truth of the theory, but we cannot deduce the theory from these sentences. The relation between facts and theories is an 'inductive' one, which means that even if we could conclusively prove the truth of the factual claims, we could not *prove* the truth of the theory. Given the above initial formulation of the criterion of meaning, this would have the consequence that even scientific theories have no cognitive meaning.

There are, obviously, different responses that could be made to this conclusion. One extreme response is to cling to the verifiability criterion as formulated and to reject the unprovable scientific theories as being meaningless. This would be like the response of a behaviourist psychologist who wishes to reject all descriptions of minds as being essentially meaningless. Some psychological theories (e.g. Freudian theories) involve reference to hidden, or unobservable, mental events. Most of our everyday descriptions of ourselves and others also appear to commit us to a belief in processes going on in our minds which are hidden from public view. The nature of these processes, as they go on in other people, is only inductively inferrable from the observations we can make. For example, we talk about beliefs, intentions, desires, subconscious fears and anxieties, thoughts, emotions, and so on, and such talk does not appear to be only about the behaviour which we can observe. The strict version of the meaning-criterion would suggest that such theorising, even on the simple level of everyday talk, is really of a metaphysical, and so meaningless, sort. The impossibility of *proving* the truth of such descriptions of mental happenings would be sufficient indication of the meaninglessness of these descriptions.

This conclusion might be objected to on the grounds that although the behaviourist would be right to 'eliminate' mentalistic terminology from his description of, and theorising about, people, if these processes were unobservable, he is wrong in thinking this to be the case. That is, our objector might well believe that strict verificationism is correct, but nevertheless disagree with the behaviourist's conclusion because one of the behaviourist's assumptions is wrong. What has been overlooked in the above argument, it could be claimed, is that even though I do not have direct (observable) knowledge of the mental life of other people, there is at least one person whose mental processes I do observe, namely myself. On this view we can still hold to the strict version of the criterion of meaning, admit that the mental processes of others are hidden from us, and still claim that mental terms are meaningful because they derive their meaning from our ability to introspect. Such introspection could give us the required proof of the correctness or incorrectness of the descriptions we use which involve reference to 'inner' processes.

Such a line of argument might seem plausible, particularly in view of the behaviourist's apparent reluctance to accept the validity of introspectively derived accounts of what goes on in our minds. However, this argument has a consequence which is as unpalatable as the behaviourist's total rejection of mentalistic description. The opponent of behaviourism has accepted the verifiability criterion of meaning – that the meaning of sentences depends upon the ways in which they are to be assessed as true or false. Another of his assumptions is that it is only in the first-person case, only in my own case, that the truth or falsity of such descriptions can be determined. Only I can introspect my own mind, and such a method of verifying what is happening inside you is not available to me. As a consequence, mental descriptions only make sense when applied to me, and do not make sense when applied to anybody else. Only I can be said to have a mind, as only I can introspect, hence verify, my thought-processes and other mental 'happenings'. Mental terminology is meaningful only when verifiable; descriptions involving reference to mental processes are verifiable only by me; hence I am the only person who can meaningfully be said to have a mind. The philosophical position called 'solipsism' results, this being the belief that I am the only real person in the world. (One is reminded of the lady who wrote to Bertrand Russell saying she was convinced of the truth of solipsism, but was distraught because she found it difficult to convince other people of its truth.)

We have suggested that both the behaviourist and solipsist have arrived at unpalatable conclusions. This *in itself* is no reason to reject the conclusions – if true premises lead, by a process of correct argument, to unpalatable conclusions, then we are not entitled to reject the conclusion simply because we dislike it. However, in this case we can re-examine our assumptions to see if they are correct, and in particular the assumption involving the strict criterion of meaning. This emphasises the importance of theories of meaning – the adoption of the criterion outlined above would rule out many other beliefs that we might otherwise have thought it rational to hold. The whole area of philosophical psychology (philosophy of mind), and psychology itself, would be revolutionised by adopting this view of meaning. It is for reasons such as this that contemporary philosophers believe philosophy of language to be of fundamental importance in most areas of philosophy. Assumptions we make about the meaning of the theories, commonsensical or abstruse, that we have about the world, can have far-reaching consequences for those theories. They can also have radical consequences for the view we might have of philosophical method and argument itself. If the verifiability criterion is correct in its claim about what are and are not meaningful sentences, then the only meaningful sentences are those which, typically, scientists use. This puts a lot of philosophers on the nonsense side of the dividing line between meaningful and meaningless discourse. The role of philosophy becomes that of analysing scientific language and codifying scientific practice. Philosophy of science becomes all the philosophy there can be.

Choosing a theory of meaning These conclusions depend, for their plausibility, on the correctness of the theory of meaning upon which they are based. How is it possible to decide whether a theory of meaning is correct or not? This is a question without an easy answer, but is obviously of major importance. It is a question that is still troubling philosophers of language, reformulated in terms of the kinds of constraints which are placed upon candidate theories of meaning. That is, what features of our language are sufficiently important for us to need to include them in a list of features that must be accounted for by any theory of meaning which is to be given serious consideration?

One such feature has already been mentioned in the previous section – our ability to understand, and communicate by using, sentences which we have never heard or used before. We do this by combining old words to form new sentences, and this combination is done systematically. This suggests that one constraint that should be placed upon proposed theories of meaning, one condition that any candidate must satisfy, is that it describes the rules governing permissible combinations. It would be hoped that these rules would also explain such facts as our ability to move from 'Harry is a short man' to 'Harry is a man'. 'Short' would be classified with the class of adjectives which contribute to the meaning of the sentences in which they appear in such a manner as to allow for the appropriate inference to be drawn. One would have to have a different rule for adjectives such as 'fake', since it would be a mistake to move from 'That is a fake diamond' to 'That is a diamond'.

This is the kind of constraint one might place upon theories of meaning. A verificationist theory could satisfy the above constraint by suggesting that the rules governing 'short' and 'fake' should take into account the conditions under which sentences containing those words are appropriately uttered. In other words, the verification-conditions of these sentences would be of major importance. The conditions which would warrant the assertion of 'Harry is a short man' would be the existence of a person, named Harry, who was a man and also short – or, more specifically, short for a man. The relevant conditions for 'That is a fake diamond' would be at least the presence of an object which looked like, but was not, a diamond. The difference between 'short' and 'fake' is explained by the fact that in the one case there is an object indicated by the noun which is qualified by the adjective, in the other case there is not. The difference in the verifying conditions would account for the different inferences which could be drawn from the two sentences.

Another constraint which a theory of meaning should satisfy is that it should not make our normal communicative ability impossible to understand, or, indeed, just plain impossible! This is the danger which faces strict verificationism, especially in its application to philosophy of mind and psychology. We have indicated that much of our everyday talk about people incorporates terms that appear to refer to publicly unobservable happenings in people's minds. Can it be plausibly maintained that such talk is nonsensical, or that we can never sensibly talk of other people's mental life? Wouldn't a more appropriate response be to claim that any theory of meaning which has this as a consequence is failing to meet the required constraints, or conditions, which such theories must satisfy in order to be deemed worth holding?

Taken by itself, the unverifiability of mental discourse could incline us one way or another. We might say that the typically mentalistic terminology which we employ – 'want', 'desire', 'hope' etc. – should not be ruled out as nonsensical, but reinterpreted. This line of thought would suggest that we are under an illusion if we think that such concepts refer to inner, private processes; that what they *really* do is describe different responses which we make to external stimuli. This is a plausible enough view for certain characteristics which we ascribe to people; somebody is brave only if he responds to dangerous situations in a certain way. However, it is not very plausible to extend this view to cover beliefs. What behaviour is to be identified with believing that $2 + 2 = 4$, or that Halley's Comet will reappear in 1986? And on this view painful experiences should be identified with the behaviour of the suffering subject, thus contradicting the belief that stoics, who completely suppress external signs of suffering, are in pain.

These uncomfortable consequences of the theory would incline many people to suggest that the theory of meaning should be adjusted, rather than to attempt a heroic reinterpretation of all of our mental vocabulary. Taken in conjunction with the other consequence – that many of our 'respectable' scientific theories would find themselves on the nonsense side of the divide – these considerations do suggest that something has gone wrong with the theory. The verifiability criterion of meaningfulness was meant to differentiate scientific sense from metaphysical nonsense, not to abolish large areas of scientific theory. Many scientific theories use concepts in their explanations which cannot be defined in terms of what is open to view – 'forces', 'currents', 'fields', 'sub-atomic particles' would all be banned by a strict application of the criterion. A relaxation of the strict version seems to be required. Instead of demanding the possibility of *proving* the truth of all meaningful sentences, we can instead recognise the meaningfulness of sentences which are only inductively related to other sentences which, in turn, would provide evidence for, rather than proof of, their truth. The looser formulation of the theory admits as meaningful all sentences for which there is evidence that would count for or against their truth.

The effect of this reformulation is immediately obvious in the philosophy of science. Instead of trying to eliminate those terms which refer to unobservable processes, or to define them in terms of what is observable, we can include them without apology in scientific explanations provided that there is some evidence to support the view that such processes are occurring. Electrons escape being written off as metaphysical entities provided that we can specify the experimental conditions which will realise their (observable) effects. But 'vital spirits' are dismissed as meaningless if their postulation doesn't make any difference to observable goings-on. The loosened criterion in effect gives backing to our 'principle of economy' which we discussed in the philosophy of science chapter (the principle asserting that the *simplest* explanation is preferable), and it does so from the perspective of a theory of meaning.

The consequences of our reformulation for philosophy of mind and psychology are less clear. We may still maintain that it is perfectly sensible to say that we feel pains, have beliefs, experience desire without *any* behavioural accompaniment. A dualist would be strongly inclined towards the view that there is no such dependency between our mental life and observable behaviour; that the evidential relationship between the two which is required even by our loosened criterion is still too tight for comfort. The dualist may allow that behaviour can be evidence for mental states, but deny that it is necessary for this evidence to be available. A defender of the criterion may admit that it is possible, in particular cases, for (e.g.) a belief never to be 'expressed' in behaviour, but argue that in general there is some such connection; that we couldn't have acquired the concept of belief unless there was such a relationship which could inform us of the occasions when it would be appropriate to ascribe beliefs to others. The meaning of the term would then, it is claimed, depend at least partly on a connection between its use and observable behaviour, all other uses being parasitic upon this central core. In individual cases we may meaningfully describe somebody as having a belief even if the belief has not been behaviourally manifested; but we could not do this unless there was, in the majority of cases, a connection between certain kinds of behaviour and holding a belief.

As we have said, our concern is not to follow up all the arguments and counter-arguments of this debate, but rather to illustrate the interconnectedness of philosophy of language with other areas in philosophy. The above discussion has illustrated this interconnectedness with respect to philosophy of science and philosophy of mind. Assumptions about which terms and sentences are meaningful, and why they are so, have had an effect upon social philosophy as well.

The particular theory of meaning which we have been discussing, the verifiability theory, was proposed partly as an explanation for the lack of progress made in metaphysical argument. The apparent impregnability of some metaphysical views is not regarded as a virtue, a sign of their supreme rationality and truth, but rather as a sign that these views are not really saying anything about the universe, and are therefore not really saying anything. It is claimed that this contrasts strongly with the state of affairs in science. Science is characterised by progressive development; its theories are constantly being modified and changed, and this is made possible by their direct relationship to evidence, which enables us to adjudicate between their truth and falsity. Since the theory of meaning proposed is a *general* theory, applying to all of language, if there are other areas of language where disputes appear to have the same character as metaphysical arguments, resisting resolution by appeal to matters of fact, then they too should be stigmatised as having the same defects, i.e. as employing cognitively meaningless language. Moral language has been put into this category.

To show how this is so, we must now explain why the qualification 'cognitively' appears when one declares a non-scientific sentence to be meaningless. One consequence of a full-blooded verificationism is that it would rule out, unless amended, large areas of literature from being meaningful. This seems to contradict linguistic practice – a great many people enjoy and appreciate poetry, even though the primary aim of a poet is not to describe and explain the world in a scientific manner. Understanding poetry would not be possible if poetry was really nonsensical, as full-blooded verificationism suggests. The response to this problem was to suggest that what poets are really doing is expressing emotions, giving external form to their emotional experiences. They hope, it is claimed, to be able to convey these experiences to others by the skilled use of language, using it in an evocative rather than a descriptive manner. This view has led verificationists to suggest that language used in this way might not be totally meaningless, but rather lacking cognitive or factual meaning: such language would not be suitable for conveying factual information about the world. Poetic discourse is possible because language can be used emotively and have 'emotive meaning'. This kind of meaning, it is held, is peculiarly apt for the expression and evocation of emotions, rather than for theoretical discussions about what exists in the world.

The verifiability criterion and social philosophy Whatever the rights and wrongs of this analysis of the language of literature, it can be easily seen that the same approach would look inviting in moral or social theory. This is another area of philosophical dispute in which there has seemed to be little chance of amicable agreement. It is true that when two people disagree about what *ought* to be done in any situation, the disagreement can sometimes be resolved by an appeal to facts. One party to the dispute may be ignorant about some of the factors relevant to it, and enlightening him may be enough to get him to change his mind. In a debate about abortion, for example, one person may express the view that abortion ought never to be performed because it might lead to the extinction of the human race through depletion of the population. Given factual information regarding the world's birth-rate, plus predictions about the likely number of abortions, were they to be legalised, this person might well change his mind.

However, there appear to be other cases which are not like this at all, where the disputants agree on all the factual information, but still disagree on the moral appraisal of it. There seems to be a fundamental disagreement of moral principle, immune to scientific arbitration. This could arise in the abortion case if one person was against all abortions on the grounds that nobody ever had the right to kill any living organism, while the opposition's contrary view was buttressed by the claim that a woman has a right to do what she likes with her body, including any living organism within it. How can this dispute be settled? If these are fundamental principles, it appears that the argument has reached a deadlock and that no resolution is possible.

An application of the verifiability criterion of meaning would suggest that the peculiarly moral terms involved, e.g. 'ought', 'good', 'bad', 'right', 'wrong', have no anchorage in the world around us. No appeal to objective reality can tell for or against the conflicting moral principles which are being upheld. A particular moral judgement, e.g. 'The execution of Joe Brown was wrong', may involve a factual element, and the disagreement about this judgement might be cleared up by examining the facts. It may be that Joe Brown was not executed, or one disputant may believe that Joe Brown committed a murder whilst another may not. However, it might also be the case that the disagreement is not about matters of fact, and it is in this case that the moral terms involved in such moral judgements would be seen, so it is claimed, to be empty of factual content. What underlies these moral evaluations are moral principles, and the verificationist would claim that these do not describe an objective reality, they serve rather to express our emotional reactions to the situations being judged. What divides the disputants would be their different emotional reactions, not anything 'objective'.

This divorce between scientific method and moral method has been commented upon by many philosophers, one of the most famous being Hume, who claimed that an 'ought' could never be derived from an 'is'. He meant that we could never derive moral judgements about what we ought to do from empirical judgements about what is the case. In more recent times, the attempt to eliminate moral statements and replace them by the statements of natural science has been termed the 'naturalistic fallacy', with the implication that any such attempt is doomed to failure. The verifiability criterion of meaning, applied to moral judgements, would suggest that although these sentences may have a factual component (that part of it which describes the action which is being morally evaluated), the more specifically moral component involves nothing more than an emotional reaction to the facts. A verificationist, then, denies that moral concepts have any factual component at all, and so, on his view, such concepts do not add anything to the cognitive meaning of the sentences in which they appear. A particularly clear statement of this view is found in A. J. Ayer's *Language, Truth and Logic.*

> Thus if I say to someone, 'You acted wrongly in stealing that money', I am not stating anything more than if I had simply said, 'You stole that money'. In adding that this action is wrong I am not making any further statement about it. I am simply evincing my moral disapproval of it. It is as if I had said, 'You stole that money', in a peculiar tone of horror, or written it with the addition of some special exclamation marks. The tone, or the exclamation marks, adds nothing to the literal meaning of the sentence. It merely serves to show that the expression of it is attended by certain feelings in the speaker. (Ayer 1936, p. 142)

This has been colloquially called the 'Boo-Hurrah' theory of ethics because it appears to reduce moral condemnation just to saying 'boo', moral praise to saying 'hurrah'. Moral disagreement isn't really disagreement at all, any more than my preference for vanilla ice-cream is a disagreement with somebody who prefers chocolate. Such a view of moral discourse was, and is, alarming to many people who dislike the apparent arbitrariness of moral judgement which it entails. One man's judgement becomes as good as another's, in so far as both are simply expressions of different emotional reactions. Moral appraisal becomes relative in the extreme – relative to each individual person. Moral relativism is, primarily, the doctrine that moral judgements are not absolutely right or wrong, that they can only be evaluated in this manner if there is some background theory, or culture, in relation to which we are evaluating the relevant behaviour. On this view 'Killing twins is wrong' is a correct moral judgement only relative to, for example, the moral code of the western world; such an evaluation would be incorrect if the judgement was about the behaviour, say, of certain tribes in Africa, which have a different code of conduct. Relativism is involved because, it would be claimed, there are no absolute standards by which to judge the rightness or wrongness of different codes and moral practices; moral evaluation is possible only in relation to the specific culture involved.

As has been mentioned, the relativism forced on us by the 'verifiability theory' of meaning is more acute. *There are no moral facts*, even of a culture-bound kind, to settle individual disputes; hence only emotive meaning is involved. The correctness or incorrectness of such 'judgements' would depend only on whether the person is giving honest expression to his emotions. That is, a moral judgement would be 'wrong', on this view, only if the individual doing the judging pretended to have a reaction that he was not really experiencing. Morality becomes relative to individuals' emotions, and moral argument is impossible.

Once again, it is not our business to pursue the subsequent discussion of this 'emotive theory of ethics'. Needless to say, the conclusions suggested by the application of the verifiability criterion of meaning have proved unacceptable to many philosophers concerned with these questions, and the debate has raged fast and furious. One reaction may be worth noting, since it throws further light on the issues involved and touches on our earlier discussion. It may be thought that the conclusion we have reached by using the verifiability criterion is valid only on a 'strict' reading of that criterion. That is, Hume may have been correct in asserting that an 'ought' can never be derived from an 'is' – but then no scientific theory can be derived, in the sense of logically deduced, from the facts either, so why the fuss? This suggests that a strategy employed for mental concepts, and other scientific terms, may be available to the defender of moral theory; he may say that just as none of our psychological terms can be exhaustively defined using only behavioural language, so none of our moral terms can be replaced by ordinary descriptive terms. This would not in itself mean that the function of moral judgements is just to express emotions; moral terms can be related to the facts by inductive links, and thereby rational argument can be restored to morality.

Such a view is enticing, but not as easy to apply in the moral case as it appeared to be in the psychological case. The position of the verificationist, loose or strict, is not that moral terms in moral judgements derive their meaning by being based on the facts; it is that the factual component of the judgement is exhausted by the non-moral (descriptive) terms in the judgement. The descriptive terms may be related to the facts, but the moral terms do not add anything to the factual component.

Compare the way 'short' and 'good' function in the sentence 'Jack is a short man' and 'Jack is a good man'. If we know what the average size of men is, then being told that Jack is short tells us that he is shorter than average. That is, the sentence will be informative to us if we have sufficient relevant background information, the information in this case being knowledge about the average height of men. What background information is relevant for interpreting 'Jack is a good man'? We would need to know, in order to know what contribution 'good' is making to the sentence, what ethical code is being presumed by the speaker. For a verificationist this boils down to knowing what kinds of actions the speaker either approves or disapproves of – which actions he likes or dislikes. 'Short' tells us something about Jack in so far as we know something about the height of men; 'good' tells us something about Jack (or his actions) only in so far as we know the speaker's likes or dislikes, his emotional reaction to Jack's behaviour. This is why a verificationist would claim that ethical terms, such as 'good', have emotive rather than factual meaning.

As we have said, this debate continues to engage philosophers, and it would take us too far afield to explore its ramifications here. Enough has been said to indicate the importance of an understanding of the phenomenon of language for a proper understanding of human beings, and consequently the energy expended by philosophers on questions concerning language should not be surprising. This chapter has looked at two questions within the general area of philosophy of language – the distinctiveness of language, and the pervasive importance of theories of meaning. In so far as language is the medium in which we think, getting clear about the first question will illuminate some of the questions which arise when we think about thinking; answers to the second problem (What is a correct theory of meaning?) assume greater importance the more we realise the interdependence between such an answer (e.g. verificationism) and the way we look at other issues in philosophy.

References and further reading

Works are referred to in the text by date of original publication, given here in brackets immediately after the author's name. Page references are to the most recent edition mentioned.

OTHER USEFUL INTRODUCTORY BOOKS

ABEL, R. (1976) *Man is the Measure* (New York: Macmillan)
AYER, A. J. (1973) *The Central Questions of Philosophy* (Harmondsworth, 1976: Penguin)
HOSPERS, J. (1967) *An Introduction to Philosophical Analysis*, 2nd ed. (London: Routledge and Kegan Paul)
RUSSELL, B. A. W. (1912) *The Problems of Philosophy* (Oxford, 1968: Oxford University Press)

CHAPTER 1

References

GLOVER, J. (1977) *Causing Death and Saving Lives* (Harmondsworth: Penguin)
GOSTIN, L. O. (1975) *A Human Condition* (London: National Association for Mental Health)
HALSBURY (1969) *Halsbury's Statutes of England*, 3rd ed., vol. 8 (London: Butterworth)
HUME, D. (1738) *A Treatise of Human Nature* (Oxford, 1968: Oxford University Press)
LAING, R. D. (1959) *The Divided Self* (Harmondsworth, 1965: Penguin)

MILL, J. S. (1859) *On Liberty*, in *Utilitarianism, On Liberty, and Considerations on Representative Government* (London, 1972: Dent)

MILL, J. S. (1861) *Utilitarianism*, ibid.

Further reading

CLARE, A. (1976) *Psychiatry in Dissent* (London: Tavistock)

FEINBERG, J. (1973) *Social Philosophy* (Englewood Cliffs, New Jersey: Prentice Hall)

FOOT, P. (1967) (ed.) *Theories of Ethics* (Oxford: Oxford University Press)

RACHELS, J. (1975) (ed.) *Moral Problems*, 2nd ed. (London: Harper & Row)

SMART, J. J. C. and WILLIAMS, B. A. O. (1973) *Utilitarianism For and Against* (Cambridge: Cambridge University Press)

These books all contain extensive bibliographies and suggestions for further reading.

Philosophy and Public Affairs, published by Princeton University Press, Princeton, New Jersey 08540, U.S.A., is a journal which specialises in articles of concern to social philosophers.

CHAPTER 2

References

DESCARTES, R. (1641) *Meditations*, in Haldane, E. S. and Ross, G. R. T. (1911) (eds and translators) *The Philosophical Works of Descartes*, vol. 1 (Cambridge, 1969: Cambridge University Press)

HOBBES, T. (1651) *Leviathan* (Harmondsworth, 1968: Penguin)

SKINNER, B. F. (1971) *Beyond Freedom and Dignity* (Harmondsworth, 1973: Penguin)

Further reading

BLACKMAN, D. (1974) *Operant Conditioning* (London: Methuen)

O'CONNOR, D. (1971) *Free Will* (London, 1972: Macmillan)

O'CONNOR, J. (1969) (ed.) *Modern Materialism: Readings on Mind-Body Identity* (New York: Harcourt, Brace & World)

RYLE, G. (1949) *The Concept of Mind* (Harmondsworth, 1963: Penguin)

SHAFFER, J. A. (1968) *Philosophy of Mind* (Englewood Cliffs, New Jersey: Prentice Hall)

CHAPTER 3

References

HANLON, J. (1974) 'Uri Geller and science', *New Scientist*, 17 October 1974, 170–85.

KIRK, G. S. and RAVEN, J. E. (1957) *The Presocratic Philosophers* (Cambridge, 1966: Cambridge University Press)

TAYLOR, J. (1976) *Superminds* (London: Pan)

VELIKOVSKY, I. (1950) *Worlds in Collision* (New York: Doubleday)

Further reading

COHEN, I. B. (1960) *The Birth of a New Physics* (London, 1961: Heinemann)

SOLOMON, J. (1973) *The Structure of Matter* (Newton Abbott: David and Charles)

SOLOMON, J. (1973) *The Structure of Space* (Newton Abbott: David and Charles)

The above three books are on the history of science.

HEMPEL, C. (1966) *Philosophy of Natural Science* (Englewood Cliffs, New Jersey: Prentice Hall)

TOULMIN, S. (1953) *The Philosophy of Science* (London: Hutchinson)

The above two books are good introductions to the philosophy of science.

RANDI, J. (1975) *The Magic of Uri Geller* (New York: Ballantine Books)

This book provides an alternative explanation of the Geller phenomenon to that provided in Taylor 1976.

There are several philosophy jourals which specialise in articles of concern to philosophers of science. We recommend the *British Journal for the Philosophy of Science,* published by Aberdeen University Press.

CHAPTER 4

References

AYER, A. J. (1936) *Language, Truth and Logic* (Harmondsworth, 1971: Penguin)

BENNETT, J. F. (1976) *Linguistic Behaviour* (Cambridge: Cambridge University Press)
BLAKEMORE, C. (1976) *Mechanics of the Mind* (Cambridge, 1977: Cambridge University Press)
BROWN, R. (1976) *A First Language* (Harmondsworth: Penguin)
DESCARTES, R. (1637) *Discourse on Method,* in Haldane, E. S. and Ross, G. R. T. (1911) (eds and translators) *The Philosophical Works of Descartes,* vol 1. (Cambridge, 1969: Cambridge University Press)
LINDEN, E. (1975) *Apes, Men, and Language* (New York, 1976: Penguin)
LOCKE, J. (1690) *An Essay Concerning Human Understanding* (Oxford, 1975: Oxford University Press)

Further reading

CHOMSKY, N. (1968) *Language and Mind* (New York: Harcourt, Brace & World)
HACKING, I. (1975) *Why Does Language Matter to Philosophy?* (Cambridge: Cambridge University Press)
HEMPEL, C. (1952) 'Problems and changes in the empiricist criterion of meaning', in Linsky, L. (1952) (ed.) *Semantics and The Philosophy of Language* (Chicago: University of Illinois Press)
LYONS, J. (1970) *Chomsky* (London: Fontana)
SEARLE, J. (1971) (ed.) *The Philosophy of Language* (Oxford: Oxford University Press)

Index